80 Years of Professor Wigner's Seminal Work "On Unitary Representations of the Inhomogeneous Lorentz Group"

80 Years of Professor Wigner's Seminal Work "On Unitary Representations of the Inhomogeneous Lorentz Group"

Editor

Julio Marny Hoff da Silva

MDPI • Basel • Beijing • Wuhan • Barcelona • Belgrade • Manchester • Tokyo • Cluj • Tianjin

Editor
Julio Marny Hoff da Silva
Universidade Estadual Paulista
Brazil

Editorial Office
MDPI
St. Alban-Anlage 66
4052 Basel, Switzerland

This is a reprint of articles from the Special Issue published online in the open access journal *Universe* (ISSN 2218-1997) (available at: https://www.mdpi.com/journal/universe/special_issues/80years).

For citation purposes, cite each article independently as indicated on the article page online and as indicated below:

LastName, A.A.; LastName, B.B.; LastName, C.C. Article Title. *Journal Name* **Year**, *Volume Number*, Page Range.

ISBN 978-3-0365-2197-8 (Hbk)
ISBN 978-3-0365-2198-5 (PDF)

Contents

universe MDPI

Editorial

Editorial to the Special Issue "80 Years of Professor Wigner's Seminal Work: On Unitary Representations of the Inhomogeneous Lorentz Group"

Julio Marny Hoff da Silva

Departamento de Física, Universidade Estadual Paulista, UNESP, Av. Dr. Ariberto Pereira da Cunha, 333, Guaratinguetá 12516-410, Brazil; julio.hoff@unesp.br

Citation: Hoff da Silva, J.M. Editorial to the Special Issue "80 Years of Professor Wigner's Seminal Work: On Unitary Representations of the Inhomogeneous Lorentz Group". *Universe* **2021**, *7*, 310. https://doi.org/10.3390/universe7080310

Received: 11 August 2021
Accepted: 13 August 2021
Published: 21 August 2021

The present Special Issue is dedicated to celebrate 80 years of the Professor Eugene Paul Wigner paper "On Unitary Representations of the Inhomogeneous Lorentz Group", published in 1939 [1]. It is almost impossible to have a fair measure of the impact of such a work. To name one of the most important achievements present in [1], the very concept of a particle is there constructed: a particle is an irreducible representation of the Poincarè group. For the contemporary reader, this assertion may sound as a kind of truism, but it certainly deserves some additional attention.

Poincarè transformations leave the Minkowski metric invariant. They may quite well be understood as symmetries of spacetime in the absence of gravity. The fulcrum in the Wigner analysis is, then, to represent the group of spacetime symmetries in the Hilbert space—the space of quantum physical states—in such a way that this representation acts also in a symmetric way, where in this last case a symmetry means a set of transformations preserving probabilities. There are many intricate problems in such an endeavor.

To begin with, it is important to recall that, although very well known, the Lorentz group is not simple. It indeed has a nontrivial topology and the task of genuine or projective representation is a problem to be addressed. Besides, continuity is also a concept generally accepted as valid, but rarely proved. These formal, but definitely important, aspects were worked out and developed in a precise manner by Wigner. Moreover, as the Lorentz group is non-compact, there is no unitary representation of finite dimension for that. Unitarity, however, is indeed a property of representations preserving norms for a connected group (as Wigner himself demonstrated in 1931 [2]).

The crucial problem related above was solved by Wigner with the help of the induced representation method, by means of which the Little Group is reached. As a typical achievement resulting from such an approach, by selecting one particle state four-momentum representative for a massive particle as $(m, \vec{0})$, the Little Group condition engenders a constraint, leading to the recognizing of $SO(3)$ as the Little Group, endowed, thus, with momentum angular algebra (even for a rest momentum as the chosen one). That is to say: the spin of a massive particle (an intrinsic angular momentum indeed) is a quantum label without which it would not be possible to represent all the spacetime symmetries in Quantum Mechanics. The particle properties are then bound to the spacetime symmetries. After 56 pages of a rigorous analysis, this is the kind of beauty arising from the Wigner analysis.

The contributions to this Special Issue may be separated out into Reviews and Research papers. In a broad-brush picture it may be reported as follows: in the paper [3] the effects of chirality in the spin entanglement between a lepton and a neutrino are investigated. The results point to an interesting framework for measuring chiral oscillations and, as a consequence, the intrinsic (bi)spinorial fermionic structure. The paper [4] deals in detail and carefully with the gauge invariance of the bosonic measure in the generating functional of chiral gauge theories, a point usually commented on, but not addressed. Going further in the Special Issue contributions, in the paper [5], the algebra of quaternions is used to achieve interesting results for the relativistic combination of non-collinear tri-velocities. As a result, the nontrivial angular dependence for such combinations is naturally absorbed in the quaternion multiplication, the same happening with other relevant results. In the paper [6], the outlines of a framework to discuss quantum gravity is presented. The ideas, comments and thoughts are clearly exposed, entailing a rich discussion. The resulting model argues that what we perceive as classical spacetime is actually the configuration space of its content.

Moving to the reviews part of the Special Issue, the review [7] brings a discussion about the special care devoted to the strongly continuous representation concept in the Wigner analysis. The relevant theorems and propositions related to the continuity concept in representation theory are reviewed in detail. The review [8] approaches the structure of gauge theories of gravity and post-Riemann geometries, discussing their recent developments and framing the analysis in the scope of understanding the very nature of spacetime and gravity in regimes where the General Relativity framework may come to debacle. The review [9] investigates the concept and properties of Dirac, Weyl and Majorana fermions in the four-dimensional Minkowski spacetime. Special attention is devoted to the quantum aspects of Weyl and Majorana fermions, discussing and clearing some misconceptions presented in the literature. The effective actions for these fermions coupled to gauge fields are reviewed in detail, and the relation between chiral and trace anomalies is extensively approached.

This Special Issue provides a well balanced mix of reviews and new results, culminating in an interesting volume celebrating and appreciating the Wigner work, but also studying physically relevant problems and methods in contemporary physics. There is rich material covering a wide range of areas, such as Quantum Mechanics and Quantum Field Theory, Mathematical Physics, and Gravitation. In some sense, the variety of subjects approached here is also a tribute to the robust foundational work of Professor Wigner: deep roots provide leafy and branched trees.

Funding: This work was supported by the Conselho Nacional de Desenvolvimento Científico CNPq (grant No. 303561/2018-1).

Institutional Review Board Statement: Not applicable.

Informed Consent Statement: Not applicable.

Data Availability Statement: Not applicable.

Acknowledgments: JMHS thanks to CNPq (grant No. 303561/2018-1) for financial support.

Conflicts of Interest: The author declares no conflict of interest.

References

1. Wigner, E.P. On Unitary Representations of the Inhomogeneous Lorentz Group. *Annals Math.* **1939**, *40*, 149. [CrossRef]
2. Wigner, E.P. *Gruppentheorie und ihre Anwendungen auf die Quantenmechanik der Atomspektren*; Springer: Braunschweig, Germany, 1931; p. 251.
3. Bittencourt, V.A.S.V.; Bernardini, A.E.; Blasone, M. Lepton-Antineutrino Entanglement and Chiral Oscillations. *Universe* **2021**, *7*, 293. [CrossRef]
4. de Lima e Silva, G.; Girardi, T.J.; Dias, S.A. On Gauge Invariance of the Bosonic Measure in Chiral Gauge Theories. *Universe* **2021** *7*, 283. [CrossRef]
5. Berry, T.; Visser, M. Relativistic Combination of Non-Collinear 3-Velocities Using Quaternions. *Universe* **2020**, *6*, 237. [CrossRef]
6. Ziaeepour, H. Making a Quantum Universe: Symmetry and Gravity. *Universe* **2020**, *6*, 194. [CrossRef]
7. Hoff da Silva, J.M.; Caires da Rocha, G.M. Strongly Continuous Representations in the Hilbert Space: A Far-Reaching Concept *Universe* **2021**, *7*, 285. [CrossRef]
8. Cabral, F.; Lobo, F.S.N.; Rubiera-Garcia, D. Fundamental Symmetries and Spacetime Geometries in Gauge Theories of Gravity—Prospects for Unified Field Theories. *Universe* **2020**, *6*, 238. [CrossRef]
9. Bonora, L.; Soldati, R.; Zalel, S. Dirac, Majorana, Weyl in 4D. *Universe* **2020**, *6*, 111. [CrossRef]

universe

MDPI

Article

Lepton-Antineutrino Entanglement and Chiral Oscillations

Victor A. S. V. Bittencourt [1,*], Alex E. Bernardini [2] and Massimo Blasone [3,4]

[1] Max Planck Institute for the Science of Light, Staudtstr. 2, PLZ 91058 Erlangen, Germany
[2] Departamento de Física, Universidade Federal de São Carlos, P.O. Box 676, São Carlos 13565-905, São Paulo, Brazil; alexeb@ufscar.br
[3] Dipartimento di Fisica, Università degli Studi di Salerno, Via Giovanni Paolo II, 132, 84084 Fisciano, Italy; mblasone@unisa.it
[4] INFN, Sezioni di Napoli, Gruppo Collegato di Salerno, 80126 Naples, Italy
* Correspondence: victor.bittencourt@mpl.mpg.de

Abstract: Dirac bispinors belong to an irreducible representation of the complete Lorentz group, which includes parity as a symmetry yielding two intrinsic discrete degrees of freedom: chirality and spin. For massive particles, chirality is not dynamically conserved, which leads to chiral oscillations. In this contribution, we describe the effects of this intrinsic structure of Dirac bispinors on the quantum entanglement encoded in a lepton-antineutrino pair. We consider that the pair is generated through weak interactions, which are intrinsically chiral, such that in the initial state the lepton and the antineutrino have definite chirality but their spins are entangled. We show that chiral oscillations induce spin entanglement oscillations and redistribute the spin entanglement to chirality-spin correlations. Such a phenomenon is prominent if the momentum of the lepton is comparable with or smaller than its mass. We further show that a Bell-like spin observable exhibits the same behavior of the spin entanglement. Such correlations do not require the knowledge of the full density matrix. Our results show novel effects of the intrinsic bispinor structure and can be used as a basis for designing experiments to probe chiral oscillations via spin correlation measurements.

Keywords: entanglement; bispinors; chirality

Citation: Bittencourt, V.A.S.V.; Bernardini, A.E.; Blasone, M. Lepton-Antineutrino Entanglement and Chiral Oscillations. *Universe* **2021**, 7, 293. https://doi.org/10.3390/universe7080293

Academic Editor: Julio Marny Hoff da Silva

Received: 13 July 2021
Accepted: 5 August 2021
Published: 9 August 2021

1. Introduction

To ensure that transformations between inertial frames preserve probabilities, Wigner proposed in his seminal work [1] that states describing particles should belong to unitary irreducible representations (irreps) of the Poincaré group. In such a framework, the degrees of freedom that a particle carries are defined by the particular irrep in question. Specifically, the irreps of the Poincaré group are classified by its Casimir invariants: momentum (mass) and the eigenvalues of the Pauli–Lubanski vector, related to the spin [2]. This description of particle states in terms of irreps of the Poincaré group has been used not only in particle and high energy physics [3] but also in connection with information theory for studying transformation properties of quantum correlations [4–7] and for quantum protocols in relativistic setups, such as clock synchronization [8,9] and teleportation [10].

Another prominent example in the context of relativistic quantum mechanics is the Dirac bispinor, which belongs to an irrep of the *complete Lorentz group*. This group includes parity as a symmetry [2], which connects the two irreps of the (proper) Lorentz group, given in terms of the two-component Weyl spinors. The irreps of the complete Lorentz group are given in terms of a $SU(2) \otimes SU(2)$ group structure, associated with two discrete degrees of freedom [2,11–13]: spin and chirality (or intrinsic parity). Since a state described by a Dirac bispinor has two dichotomic degrees of freedom, it can be understood as a two-qubit state [11,12]. In general, chirality and spin of a single particle can become entangled under external potentials [14], as noticed for Dirac-like systems, such as bilayer graphene [15].

The intrinsic structure of the Dirac bispinor has also implications for the dynamics of free particles. Helicity (the projection of spin in the direction of the momentum) is a conserved quantity, but chirality is not [16]. In fact, the different chiral components of a bispinor are coupled via the mass term of the Dirac Hamiltonian. This generates chiral oscillations [17], which can be related to the *Zitterbewegung* effect [18]. The dynamical features of chiral oscillations are particularly relevant for dynamics of neutrinos [19–21]. Neutrinos and antineutrinos only interact via weak processes, which are inherently chiral [22] and create such particles in states with definite chirality. Under free dynamics, such massive (anti)neutrino states exhibit a finite probability of being in the opposite chirality. The amplitude of these oscillations depends on the ratio between mass and energy and is prominent in the non-relativistic dynamical regime [23]. Since only one chiral component is measurable through weak processes, chiral oscillations can yield a depletion of the measured flux of the cosmic neutrino background [24–26].

In this contribution, we show that chiral oscillations affect quantum correlations shared between spins of a lepton-antineutrino pair. We consider a singlet state of a lepton and a massive antineutrino propagating along opposite directions. Assuming that such state is created by weak interactions (e.g., by the decay of a pion, as seen in the pion's rest frame), the massive antineutrino and the lepton have, at $t = 0$, definite chiralities, modeled here by a chiral projection of the Dirac bispinors associated with the antineutrino and with the lepton. The spins are in a perfect anticorrelated spin state, as depicted in Figure 1. Under free evolution, chiral oscillations change the superposition form of the initial state, inducing oscillations on the correlations shared by the spins. We compute such oscillations both for the quantum entanglement encoded in the spins [27] and for Bell-type spin correlations [28].

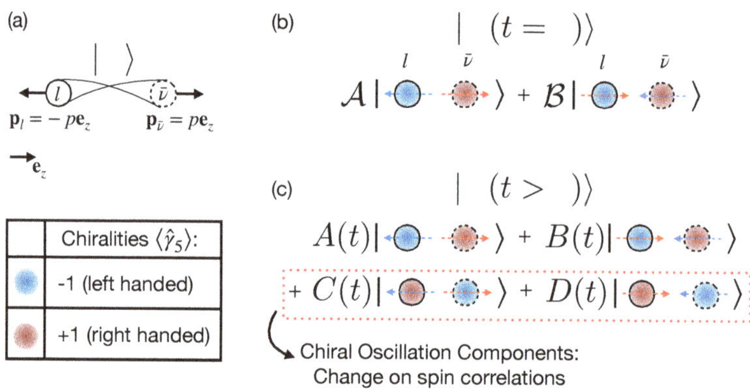

Figure 1. Framework: (**a**) a lepton and an antineutrino propagate in opposite directions with same momentum. Each is described by a Dirac bispinor, carrying spin and chirality degrees of freedom. (**b**) At $t = 0$, the state is a superposition in which the lepton and the antineutrino have definite chiralities. The coefficients $\mathcal{A}$ and $\mathcal{B}$ depend on the masses of the particles and on the momentum. (**c**) Under free evolution, the chiralities oscillate, changing the initial superposition and the correlations shared between the spins.

We show that chiral oscillations induce a redistribution of entanglement from spin-spin to spin-chirality. As a consequence, the amount of entanglement encoded only on the spins of the pair oscillates in time with a frequency that depends on the dynamical regime of propagation of both particles. In fact, even in the configuration where the lepton mass is much larger than the antineutrino mass, oscillations of the spin-spin entanglement are still relevant provided that the lepton is in the non-relativistic dynamical regime. Such entanglement oscillations exhibit a maximum amplitude for momenta of the order of the antineutrino mass, a resonance-like behavior. The spin-spin Bell-type correlation exhibits the same behavior of the spin-spin entanglement, but since its measurement requires only specific spin-spin correlations (and not a full reconstruction of the density matrix), it is more convenient for experiments.

2. Chiral Oscillations and Spin-Spin Entanglement

We consider the dynamics of a state describing a lepton-antineutrino pair propagating in opposite directions with the same momentum. We assume that the lepton (hereafter indicated by the subscript l) with mass m_l has momentum $-p\mathbf{e}_z$, while the antineutrino (indicated by the subscript $\bar{\nu}$) with mass $m_{\bar{\nu}}$ has momentum $p\mathbf{e}_z$, as depicted in Figure 1. As a toy model for the creation process, we assume the following superposition

$$|\Phi\rangle = \frac{|v_\uparrow(p, m_{\bar{\nu}})\rangle \otimes |u_\downarrow(-p, m_l)\rangle - |v_\downarrow(p, m_{\bar{\nu}})\rangle \otimes |u_\uparrow(-p, m_l)\rangle}{\sqrt{2}} \tag{1}$$

where $|u_{\uparrow(\downarrow)}(p,m)\rangle$ and $|v_{\uparrow(\downarrow)}(p,m)\rangle$ denote the Dirac bispinors with parallel ($\uparrow$) and antiparallel ($\downarrow$) spins polarized along $\mathbf{e}_z$. Those bispinors are the eigenstates of the free Dirac Hamiltonian

$$\hat{H} = \hat{\mathbf{p}} \cdot \hat{\boldsymbol{\alpha}} + m\hat{\beta}, \tag{2}$$

with energies $\pm E_{p,m} = \pm\sqrt{p^2 + m^2}$, where neutral units $c = \hbar = 1$ have been considered, boldface letters indicate vectors, and hats "^" denote operators. The 4×4 Dirac matrices $\hat{\alpha}$ and $\hat{\beta}$ satisfy the anti-commutation relations $\hat{\alpha}_i\hat{\alpha}_j + \hat{\alpha}_j\hat{\alpha}_i = 2\delta_{ij}\hat{I}_4$, $\hat{\alpha}_i\hat{\beta} + \hat{\beta}\hat{\alpha}_i = 0$, and $\hat{\beta}^2 = \hat{I}_4$, where $\hat{I}_N$ denotes the $N \times N$ identity matrix. We adopt the chiral representation of the Dirac matrices [29]

$$\hat{\alpha}_i = \begin{bmatrix} \hat{\sigma}_i & 0 \\ 0 & -\hat{\sigma}_i \end{bmatrix}, \qquad \hat{\beta} = \begin{bmatrix} 0 & \hat{I}_2 \\ \hat{I}_2 & 0 \end{bmatrix}, \tag{3}$$

with $\hat{\sigma}_i$ denoting the Pauli matrices, which returns for the bispinors

$$\begin{aligned} |u_\uparrow(p,m)\rangle &= N_{p,m}\begin{bmatrix} f_+(p,m)|\uparrow\rangle \\ f_-(p,m)|\uparrow\rangle \end{bmatrix}, & |u_\downarrow(p,m)\rangle &= N_{p,m}\begin{bmatrix} f_-(p,m)|\downarrow\rangle \\ f_+(p,m)|\downarrow\rangle \end{bmatrix}, \\ |v_\uparrow(p,m)\rangle &= N_{p,m}\begin{bmatrix} f_+(p,m)|\uparrow\rangle \\ -f_-(p,m)|\uparrow\rangle \end{bmatrix}, & |v_\downarrow(p,m)\rangle &= N_{p,m}\begin{bmatrix} f_-(p,m)|\downarrow\rangle \\ -f_+(p,m)|\downarrow\rangle \end{bmatrix}, \end{aligned} \tag{4}$$

where $|\uparrow\rangle$ and $|\downarrow\rangle$ are the eigenstates of $\hat{\sigma}_z$, and the short hand notation sets

$$\begin{aligned} N_{p,m} &= \sqrt{\frac{E_{p,m} + m}{4E_{p,m}}} \\ f_\pm(p,m) &= 1 \pm \frac{p}{E_{p,m} + m}. \end{aligned} \tag{5}$$

Finally, the normalization of the bispinors is $\langle u_s(p,m)|u_l(p,m)\rangle = \langle v_s(p,m)|v_l(p,m)\rangle = \delta_{sl}$, and the orthogonality relations are $\langle u_s(p,m)|v_l(-p,m)\rangle = 0$.

The choice of the singlet state (1) is motivated by the decay of a pion into a lepton-antineutrino pair in the center of mass of the pion [30–32], a process generated by weak interactions. Since weak interactions are inherently chiral and select definite chiral components of the bispinors, we model such an effect by projecting the state (1) into definite chiralities: the antineutrino into right (positive) chirality and the lepton into left (negative) chirality, such that, at $t = 0$, the state is given by

$$|\Psi(0)\rangle = \frac{\hat{\Pi}_R^{(\bar{\nu})} \otimes \hat{\Pi}_L^{(l)}|\Phi\rangle}{\langle\Phi|\hat{\Pi}_R^{(\bar{\nu})} \otimes \hat{\Pi}_L^{(l)}|\Phi\rangle}. \tag{6}$$

The chirality projectors are given in terms of the chiral matrix $\hat{\gamma}_5 = \mathrm{diag}[\hat{I}_2, -\hat{I}_2]$ by

$$\hat{\Pi}_{R(L)}^{(A)} = \frac{\hat{I}^{(A)} + (-)\hat{\gamma}_5^{(A)}}{2}, \quad (A = \bar{\nu}, l) \tag{7}$$

such that

$$|\Psi(0)\rangle = \mathcal{A}(p, m_l, m_{\bar{\nu}})|\bar{\nu}_\uparrow(0)\rangle \otimes |l_\downarrow(0)\rangle - \mathcal{B}(p, m_l, m_{\bar{\nu}})|\bar{\nu}_\downarrow(0)\rangle \otimes |l_\uparrow(0)\rangle. \tag{8}$$

The chirality projected states at $t = 0$ are

$$|\bar{\nu}_{\uparrow(\downarrow)}(0)\rangle = \begin{bmatrix} |\uparrow(\downarrow)\rangle \\ 0 \end{bmatrix}, \quad |l_{\uparrow(\downarrow)}(0)\rangle = \begin{bmatrix} 0 \\ |\uparrow(\downarrow)\rangle \end{bmatrix}, \tag{9}$$

and the coefficients of the superposition are given by

$$\begin{aligned} \mathcal{A}(p, m_l, m_{\bar{\nu}}) &= N_{p,m_l} N_{p,m_{\bar{\nu}}} f_+(p, m_{\bar{\nu}}) f_-(p, m_l) \left[\frac{1}{2} - \frac{p^2}{2E_{p,m_l}E_{p,m_{\bar{\nu}}}}\right]^{-\frac{1}{2}}, \\ \mathcal{B}(p, m_l, m_{\bar{\nu}}) &= N_{p,m_l} N_{p,m_{\bar{\nu}}} f_-(p, m_{\bar{\nu}}) f_+(p, m_l) \left[\frac{1}{2} - \frac{p^2}{2E_{p,m_l}E_{p,m_{\bar{\nu}}}}\right]^{-\frac{1}{2}}. \end{aligned} \tag{10}$$

If $p \gg m_{\bar{\nu}}$, then $\mathcal{A} \gg \mathcal{B}$, and the largest contribution to the superposition (8) is the first term.

Dirac bispinors belong to an irreducible representation of the complete Lorentz group, and as such they carry two intrinsic dichotomic degrees of freedom [2]: chirality (or intrinsic parity) and spin. The former degree of freedom is related to the inclusion of spatial parity as a symmetry. In this framework, we can understand each single particle state as a two qubit state [11], and a bispinor state belongs to a composite Hilbert space $\mathcal{H}_S \otimes \mathcal{H}_C$ with $\dim[\mathcal{H}_C] = \dim[\mathcal{H}_S] = 2$. Here, $\mathcal{H}_{S(C)}$ is associated with the spin (chirality) degree of freedom. Therefore, a two particle state, such as (1), can be interpreted as a four qubit state [13]. The quantum bits are chirality and spin of the antineutrino ($C_{\bar{\nu}}$ and $S_{\bar{\nu}}$, respectively) and chirality and spin of the lepton (C_l and S_l, respectively). The state of the lepton-antineutrino pair is then described in the composite Hilbert space $\mathcal{H}_{C_{\bar{\nu}}} \otimes \mathcal{H}_{S_{\bar{\nu}}} \otimes \mathcal{H}_{C_l} \otimes \mathcal{H}_{S_l}$.

The superposition (1) is an entangled state: it can not be written as the tensor product of pure states $|\psi_{C_{\bar{\nu}}}\rangle \otimes |\psi_{S_{\bar{\nu}}}\rangle \otimes |\psi_{C_l}\rangle \otimes |\psi_{S_l}\rangle$, where $|\psi_A\rangle \in \mathcal{H}_A$ with A denoting the degrees of freedom. The entanglement encoded in (1) involves all four degrees of freedom of the state. On the other hand, the chiral projected state (8), describing the intrinsic chiral character of weak interactions, encodes entanglement only between the spins. In fact, we can readily write $|\Psi(0)\rangle = |+_{C_{\bar{\nu}}}\rangle \otimes |-_{C_l}\rangle \otimes |\Psi_{S_{\bar{\nu}},S_l}\rangle$, with $|\pm_A\rangle$ denoting the positive (negative) chirality of $A = C_{\bar{\nu},l}$, and

$$|\Psi_{S_{\bar{\nu}},S_l}\rangle = \mathcal{A}(p, m_l, m_{\bar{\nu}})|\uparrow_{S_{\bar{\nu}}}\rangle \otimes |\downarrow_{S_l}\rangle - \mathcal{B}(p, m_l, m_{\bar{\nu}})|\downarrow_{S_{\bar{\nu}}}\rangle \otimes |\uparrow_{S_l}\rangle \tag{11}$$

is the joint spin state at $t = 0$.

To compute the amount of entanglement shared between the spins, we adopt the entanglement quantifier called *negativity* [33,34]. The Peres separability criterion [33] states that the partial transposed density matrix of a separable state has only positive eigenvalues For a two-qubit state, partial transposition corresponds to the transformation $|ij\rangle\langle kl| \to |il\rangle\langle kj|$, i.e., it is a transposition on the subspace of only one of the subsystems. The criterion allows the definition of the negativity as an entanglement quantifier for two-qubit states given a density matrix ϱ representing the state, it is defined as $\mathcal{N}_{S_{\bar{\nu}},S_l}[\varrho] = ||\varrho^T|| - 1$, where $||\varrho^T||$ is the trace norm of the partial transposed matrix ϱ^T, given in terms of its eigenvalues λ_i by $||\varrho^T|| = \sum_i |\lambda_i|$. This entanglement quantifier is valid for both pure and mixed states [27].

We can now calculate the entanglement shared between the spins by first evaluating the spin-reduced density matrix:

$$\begin{aligned}\rho_{S_{\bar{\nu}},S_l}(0) &= \mathrm{Tr}_{C_l,C_{\bar{\nu}}}[|\Psi(0)\rangle\langle\Psi(0)|] = |\Psi_{S_{\bar{\nu}},S_l}\rangle\langle\Psi_{S_{\bar{\nu}},S_l}| \\ &= \mathcal{A}^2(p,m_l,m_{\bar{\nu}})|\uparrow_{\bar{\nu}}\downarrow_l\rangle\langle\uparrow_{\bar{\nu}}\downarrow_l| + \mathcal{B}^2(p,m_l,m_{\bar{\nu}})|\downarrow_{\bar{\nu}}\uparrow_l\rangle\langle\downarrow_{\bar{\nu}}\uparrow_l| \\ &- \mathcal{A}(p,m_l,m_{\bar{\nu}})\mathcal{B}(p,m_l,m_{\bar{\nu}})\Big[|\uparrow_{\bar{\nu}}\downarrow_l\rangle\langle\downarrow_{\bar{\nu}}\uparrow_l| + |\downarrow_{\bar{\nu}}\uparrow_l\rangle\langle\uparrow_{\bar{\nu}}\downarrow_l|\Big].\end{aligned} \tag{12}$$

Partial transposition yields $\rho^T_{S_{\bar{\nu}},S_l}$ from which we obtain the spin-spin negativity for the state at $t = 0$

$$\mathcal{N}_{S_{\bar{\nu}},S_l}(0) \equiv \mathcal{N}[\rho_{S_{\bar{\nu}},S_l}(0)] = 2|\mathcal{A}(p,m_l,m_{\bar{\nu}})\mathcal{B}(p,m_l,m_{\bar{\nu}})|. \tag{13}$$

Figure 2 depicts $\mathcal{N}_{S_l,S_{\bar{\nu}}}(0)$ as a function of the momentum p and the lepton mass m_l in units of the antineutrino mass $m_{\bar{\nu}}$. The amount of entanglement depends both on the dynamical regime and on the lepton-antineutrino mass ratio. For $m_l \gg m_{\bar{\nu}}$ and $p \gg m_{l,\bar{\nu}}$, the spin-spin entanglement vanishes. In this limit, the antineutrino is ultra-relativistic, and thus its chirality equals its helicity. In terms of (8), $\mathcal{A} \gg \mathcal{B}$ in this limit, and the state becomes separable. In a pion decay process, this is expected due to the small mass of antineutrinos. One can also understand the separability of the state in terms of conservation of angular momentum: the initial pion has zero spin; therefore, since ultra-relativistic neutrinos have a definite helicity, the spin-polarization of the lepton is fixed [30]. Furthermore, the entanglement between spins is more prominent for $m_\nu > p$, even if $m_{\bar{\nu}} \ll m_l$.

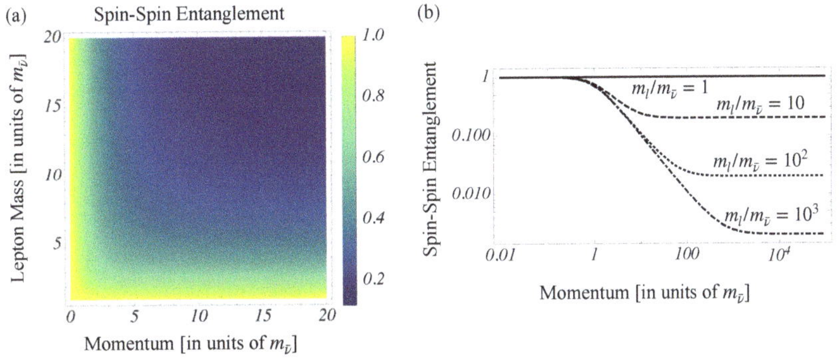

Figure 2. (**a**) Entanglement between the lepton and the antineutrino spins in the initial state (8) as a function of the momentum and of the lepton mass (in units of the antineutrino mass). (**b**) Same as (**a**) but for specific lepton-antineutrino mass ratios as a function of the momentum (in log-log scale).

Since chirality is not a conserved quantity under the free Dirac equation, temporal evolution can induce chiral oscillations [17,18]. For a two-particle state such as (8), this yields a change in the form of the superposition that dynamically redistributes the quantum correlations, encoded initially only between the spins, to other partitions of the system. Thus, because both particles are massive and described by Dirac bispinors, time evolution induces entanglement oscillations whose characteristics are intrinsically linked to those of the chiral oscillations.

For any given bispinor $|w(0)\rangle$, its temporal evolution can be obtained by a decomposition into the Dirac bispinors as [29,35]

$$|w(t)\rangle = \sum_{s=\uparrow,\downarrow} U_{w,s}e^{-iE_{p,m}t}|u_s(p,m)\rangle + V_{w,s}e^{iE_{p,m}t}|v_s(-p,m)\rangle, \tag{14}$$

where $U_{w,s} = \langle u_s(p,m)|w(0)\rangle$ and $V_{w,s} = \langle v_s(-p,m)|w(0)\rangle$. For the joint lepton-antineutrino state, we obtain

$$|\Psi(t)\rangle = \mathcal{A}(p,m_l,m_{\bar{\nu}})|\bar{\nu}_\uparrow(t)\rangle \otimes |l_\downarrow(t)\rangle - \mathcal{B}(p,m_l,m_{\bar{\nu}})|\bar{\nu}_\downarrow(t)\rangle \otimes |l_\uparrow(t)\rangle, \tag{15}$$

where the antineutrino components are given by

$$\begin{aligned} |\bar{\nu}_\uparrow(t)\rangle &= \mathcal{N}_{p,m_{\bar{\nu}}}\Big[e^{-iE_{p,m_{\bar{\nu}}}t}f_+(p,m_{\bar{\nu}})|u_\uparrow(p,m_{\bar{\nu}})\rangle + e^{iE_{p,m_{\bar{\nu}}}t}f_-(p,m_{\bar{\nu}})|v_\uparrow(-p,m_{\bar{\nu}})\rangle\Big], \\ |\bar{\nu}_\downarrow(t)\rangle &= \mathcal{N}_{p,m_{\bar{\nu}}}\Big[e^{-iE_{p,m_{\bar{\nu}}}t}f_-(p,m_{\bar{\nu}})|u_\downarrow(p,m_{\bar{\nu}})\rangle + e^{iE_{p,m_{\bar{\nu}}}t}f_+(p,m_{\bar{\nu}})|v_\downarrow(-p,m_{\bar{\nu}})\rangle\Big], \end{aligned} \tag{16}$$

and the lepton components are

$$\begin{aligned} |l_\uparrow(t)\rangle &= \mathcal{N}_{p,m_l}\Big[e^{-iE_{p,m_l}t}f_+(p,m_l)|u_\uparrow(-p,m_l)\rangle - e^{iE_{p,m_l}t}f_-(p,m_l)|v_\uparrow(p,m_l)\rangle\Big], \\ |l_\downarrow(t)\rangle &= \mathcal{N}_{p,m_l}\Big[e^{-iE_{p,m_l}t}f_-(p,m_l)|u_\downarrow(-p,m_l)\rangle - e^{iE_{p,m_l}t}f_+(p,m_l)|v_\downarrow(p,m_l)\rangle\Big]. \end{aligned} \tag{17}$$

With the time-evolved state (15), we can now compute the quantities of interest: the spin-spin entanglement and the average chiralities of the lepton and of the antineutrino. Since $\langle\bar{\nu}_i(t)|\bar{\nu}_j(t)\rangle = \langle l_i(t)|l_j(t)\rangle = \delta_{ij}$ ($\{i,j\} =\uparrow,\downarrow$), we obtain the reduced density matrices for the antineutrino and the lepton by

$$\begin{aligned} \rho_{\bar{\nu}}(t) &= \mathrm{Tr}_l[|\Psi(t)\rangle\langle\Psi(t)|] \\ &= \mathcal{A}^2(p,m_l,m_{\bar{\nu}})|\bar{\nu}_\uparrow(t)\rangle\langle\bar{\nu}_\uparrow(t)| + \mathcal{B}^2(p,m_l,m_{\bar{\nu}})|\bar{\nu}_\downarrow(t)\rangle\langle\bar{\nu}_\downarrow(t)|, \\ \rho_l(t) &= \mathrm{Tr}_{\bar{\nu}}[|\Psi(t)\rangle\langle\Psi(t)|] \\ &= \mathcal{A}^2(p,m_l,m_{\bar{\nu}})|l_\downarrow(t)\rangle\langle l_\downarrow(t)| + \mathcal{B}^2(p,m_l,m_{\bar{\nu}})|l_\uparrow(t)\rangle\langle l_\uparrow(t)|. \end{aligned} \tag{18}$$

We first notice that $\rho_{\bar{\nu},l}(t)$ are mixed states:

$$\begin{aligned} \mathrm{Tr}[\rho^2_{\bar{\nu}}(t)] = \mathrm{Tr}[\rho^2_l(t)] &= \mathcal{A}^4(p,m_l,m_{\bar{\nu}}) + \mathcal{B}^4(p,m_l,m_{\bar{\nu}}) \\ &= 1 - \frac{\mathcal{N}^2_{S_{\bar{\nu}},S_l}(0)}{2} < 1. \end{aligned} \tag{19}$$

Since the evolution is unitary, the degree of mixedness $\mathrm{Tr}[\rho^2_{\bar{\nu},l}(t)]$ is time-independent. The joint state (15) is pure at all times; therefore, (19) computes the total amount of entanglement encoded in the bipartition $(C_{\bar{\nu}}, S_{\bar{\nu}}); (C_l, S_l)$ [36], that is, the entanglement between all the degrees of freedom of the antineutrino as a whole and all the degrees of freedom of the lepton as a whole. Such entanglement is constant in time and given in terms of the initial spin-spin entanglement via the term $\propto \mathcal{N}_{S_{\bar{\nu}},S_l}(0)$ (see also Equation (13)). The average chiralities of the antineutrino and lepton are given by $\langle\hat{\gamma}_5\rangle_A(t) = \mathrm{Tr}_A[\hat{\gamma}_5^{(A)}\rho_A(t)]$ with $A = \bar{\nu}, l$ and read

$$\begin{aligned} \langle\hat{\gamma}_5\rangle_{\bar{\nu}}(t) &= 1 - \frac{m^2_{\bar{\nu}}}{E^2_{p,m_{\bar{\nu}}}}\big[1 - \cos(2E_{p,m_{\bar{\nu}}}t)\big], \\ \langle\hat{\gamma}_5\rangle_l(t) &= -1 + \frac{m^2_l}{E^2_{p,m_l}}\big[1 - \cos(2E_{p,m_l}t)\big]. \end{aligned} \tag{20}$$

Different from (8), it is not possible to write $|\Psi(t)\rangle = |\psi_{C_{\bar{\nu}}}(t)\rangle \otimes |\psi_{C_l}(t)\rangle \otimes |\psi_{S_{\bar{\nu}},S_l}(t)\rangle$, that is, the chiralities and the spins become entangled. The free evolution under the Dirac equation induces oscillations between left- and right-handed chiralities for both antineutrino and lepton, which changes the initial superposition and redistributes the correlation content carried by the state. The density matrix of the spins $\rho_{S_{\bar{\nu}},S_l}(t) = \mathrm{Tr}_{\mathrm{Chirality}}[|\Psi(t)\rangle\langle\Psi(t)|]$

is a mixed state with entanglement dynamics directly affected by chiral oscillations. The entanglement between the spins is again evaluated in terms of the negativity:

$$\mathcal{N}_{S_{\bar{\nu}},S_l}(t) \equiv \mathcal{N}[\rho_{S_{\bar{\nu}},S_l}(t)] = ||\rho^T_{S_{\bar{\nu}},S_l}(t)|| - 1 = \mathcal{N}_{S_{\bar{\nu}},S_l}(0)\Gamma(t) \tag{21}$$

with the time-dependent factor given in terms of the chiralities (20) as

$$\Gamma(t) = \sum_{j=\bar{\nu},l}\left[1 - \frac{p^2}{m_j^2}\left(\langle\hat{\gamma}_5\rangle_j(t) - 1\right)^2\right]^{\frac{1}{2}}. \tag{22}$$

The degree of mixedness of the spin density matrix reads

$$\mathrm{Tr}[\rho^2_{S_{\bar{\nu}},S_l}(t)] = 1 - \frac{\mathcal{N}^2_{S_{\bar{\nu}},S_l}(0)(1 - |\Gamma(t)|^2)}{2}, \tag{23}$$

which quantifies the entanglement in the bipartition $(S_{\bar{\nu}}, S_l); (C_{\bar{\nu}}, C_l)$, that is, the entanglement between spins and chiralities. The fact that $\mathrm{Tr}[\rho^2_{S_{\bar{\nu}},S_l}(t)] < 1$ indicates that the entanglement initially encoded only between the spins redistributes into spin-chirality entanglement.

Figure 3 depicts the average chirality of the lepton (a), of the antineutrino (b), and the spin-spin entanglement (c) as a function of the momentum p and of the time. The spin-spin entanglement oscillations depend on the chiral oscillations of both antineutrino and lepton. In the limit $m_{\bar{\nu}} \ll p$, the spin-spin entanglement vanishes, since the antineutrino has definite helicity and chirality. Furthermore, at intermediate dynamical regimes, the entanglement exhibits two oscillation frequencies: one related to the chiral oscillations of the antineutrino, and the other to the chiral oscillations of the lepton. For $p < m_{\bar{\nu}}$, the amplitude of the entanglement oscillations is suppressed, and the spin-spin entanglement is robust to the chiral oscillations. We therefore notice that the entanglement oscillations exhibit a resonance-like behavior as a function of the momentum: the amplitude of entanglement oscillations are enhanced in the dynamical regime $p \sim m_{\bar{\nu}}$.

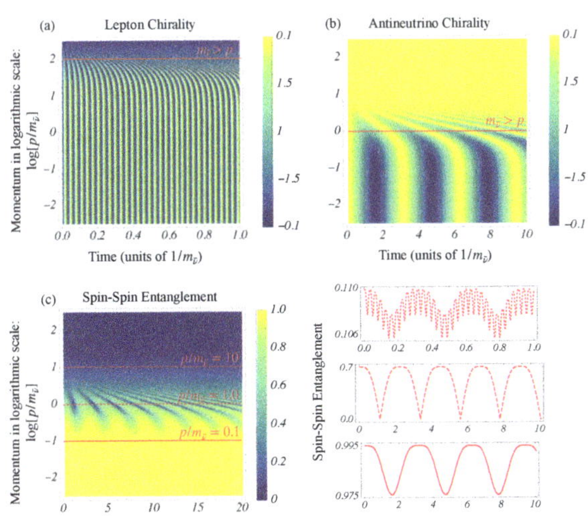

Figure 3. (**a**) Average lepton chirality, (**b**) average antineutrino chirality, and (**c**) spin-spin entanglement as a function of the momentum (in units of the antineutrino mass and in log scale) and of time. In (**a**,**b**) for $m_{l,\bar{\nu}} < p$, chiral oscillations are suppressed. Correspondingly, the spin-spin entanglement exhibits a behavior similar to the antineutrino chirality: for $m_{\bar{\nu}} \ll p$, entanglement is suppressed. Results for $m_l/m_{\bar{\nu}} = 10^2$.

3. Chiral Oscillations on Spin Correlations

Chiral oscillations impact the spin-spin entanglement shared between the antineutrino and the lepton. Nevertheless, the measurement of such correlations requires the full knowledge of the density matrix of the state, a task only accomplished by a tomographic reconstruction of the state.

A different approach to the problem consists of the measurement of Bell spin correlations. Since the spin-spin entanglement is modified by chiral oscillations, we expect that joint spin observables are also influenced by the chirality dynamics. In particular, we consider the following quantity:

$$B[\rho(t)] = |\langle \hat{S}_{\bar{\nu},1} \otimes \hat{S}_{l,1} \rangle + \langle \hat{S}_{\bar{\nu},1} \otimes \hat{S}_{l,2} \rangle + \langle \hat{S}_{\bar{\nu},2} \otimes \hat{S}_{l,1} \rangle - \langle \hat{S}_{\bar{\nu},2} \otimes \hat{S}_{l,2} \rangle|, \tag{24}$$

which is the Bell observable that was first proposed to investigate non-local correlations [28]. Considering

$$\begin{aligned} \hat{S}_{\bar{\nu},1} &= \hat{S}_{\bar{\nu},x}, \qquad \hat{S}_{l,1} = -\frac{\hat{S}_{l,x} + S_{l,y}}{\sqrt{2}} \\ \hat{S}_{\bar{\nu},2} &= \hat{S}_{\bar{\nu},y}, \qquad \hat{S}_{l,2} = \frac{-\hat{S}_{l,x} + S_{l,y}}{\sqrt{2}}, \end{aligned} \tag{25}$$

for pure states, $B[\rho] > 2$ indicates that the correlations shared between the spins are non-local and that the state is entangled. We notice, however, that even though for $t > 0$, the spin-spin reduced density matrix is not a pure state, $B[\rho]$ still quantifies spin correlations that are affected by chiral oscillations. To quantify the total amount of non-local correlations, which does not coincide with entanglement for mixed states, one has to maximize the quantity B over all possible sets of spin operators $\{\hat{S}_{\bar{\nu},i}, \hat{S}_{l,i}\}$ [28]. We instead consider the simpler framework of correlations measured with the set (25). Such a type of correlations can be measured via, e.g., Stern–Gerlach apparatuses.

Figure 4 shows $B[\rho(t)]$ for the time-evolved state (15) as a function of the momentum and of time. The observable exhibits an oscillatory behavior similar to the spin-spin entanglement (c.f. Figure 3). Nevertheless, we notice that at intermediate dynamical scales (where the antineutrino is ultra-relativistic but the lepton is not), there are no clear beatings as was observed for the entanglement. Nevertheless, the imprints of chiral oscillations in B are still prominent. Additionally, the Bell observable exhibits the same enhancement of oscillation amplitudes obtained for the spin-spin entanglement when the mass of the antineutrino is comparable with the momentum.

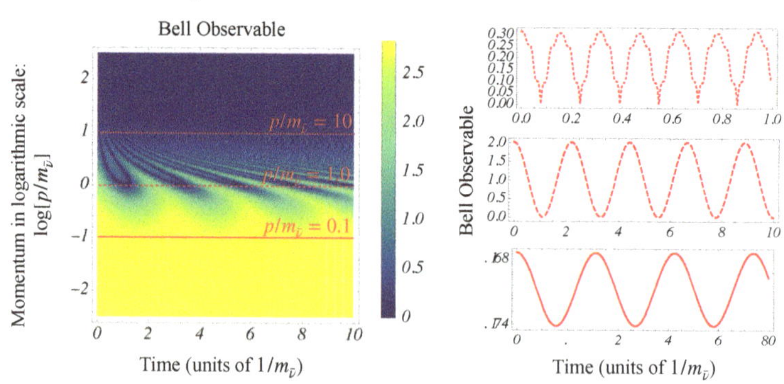

Figure 4. Bell observable (24) as a function of the momentum (in units of the antineutrino mass and in log scale) and of time for $m_l/m_{\bar{\nu}} = 10^2$.

4. Conclusions

In summary, we have discussed the effects of chirality, a degree of freedom intrinsic to massive fermions described by the Dirac equation, in the spin entanglement between a lepton and an antineutrino. We have considered a prototypical state modeling, for example, the generation of the pair by a charged meson decay (for example, a pion). To take into account the intrinsic chiral character of weak interactions, we have further considered a chirality projection into the initial superposition.

While helicity is conserved under the Dirac equation dynamics, chirality is not. We have shown that chiral oscillations change the form of the superposition of the pair and have an imprint on the entanglement shared between the spin of the lepton and the spin of the antineutrino. The entanglement dynamics is more prominent in the case where both particles are non-relativistic, even though at intermediate dynamical regimes, the spin-spin entanglement exhibits oscillations related to both antineutrino and lepton chiral oscillations. In fact, the entanglement oscillations exhibit a resonance-like behavior with the mass of the antineutrino. Finally, we have also described how chiral oscillations are present in spin-spin correlations as measured by a Bell-like observable, where features due to chiral oscillations are present. Although we considered here a simpler scenario in which only one generation of lepton is taken into account, flavor mixing intrinsic to the dynamics of massive neutrinos can be included at the level of Dirac bispinors by following [19,20]. In this case, we expect a further oscillation scale related to the flavor oscillations [23] and the generation of correlations with the flavor degree of freedom [37,38]. A further step is the inclusion of wave-packets, which also influence flavor oscillations. Finally, the formalism adopted in this paper is that of single-particle relativistic quantum mechanics. A correct description of neutrino dynamics and, in particular, the inclusion of flavor oscillations requires quantum field theory [39,40].

Our results indicate a novel framework for measuring chiral oscillations, and thus the intrinsic bispinor structure of fermions, through the dynamics of either quantum entanglement, which requires a full tomography of the spin density matrix, or by spin correlations. It should be noticed that in typical pion decay experiments (for example into a muon and a muon antineutrino), the antineutrino is typically ultra-relativistic [30], and the effects reported here should be very small. Another possibility is the generation of the entangled state via scattering processes, allowing the observation of the pair in the intermediate dynamical regime, which have more prominent imprints of chirality. Finally, even though here we considered the case of an antineutrino entangled with a fermion, the framework can be readily adapted to describe chiral oscillation effects on the dynamics of two-fermion states, including fermion-antifermion pairs. This can be particularly relevant for the study of quantum correlations and entanglement in particle physics, a growing research field [41–46].

This work is a testimony of the legacy of Wigner's seminal work, which comprises several topics of physics and mathematics: from group structures and their representations to the consequences of describing quantum states with irreps of specific groups and their possible effects in physical systems.

Author Contributions: Conceptualization, V.A.S.V.B., A.E.B. and M.B.; formal analysis, V.A.S.V.B.; Writing—original draft, review and editing, V.A.S.V.B., A.E.B. and M.B. All authors have read and agreed to the published version of the manuscript.

Funding: V.A.S.V.B. acknowledges financial support from the Max Planck Gesellschaft through an Independent Max Planck Research Group.

Institutional Review Board Statement: Not Applicable.

Informed Consent Statement: Not Applicable.

Data Availability Statement: All plots presented in this paper were done using the formulas derived in the text with the corresponding parameters showed in the plot legends.

Conflicts of Interest: The authors declare no conflict of interest.

References

1. Wigner, E. On unitary representations of the inhomogeneous Lorentz Group. *Ann. Math.* **1939**, *40*, 149. [CrossRef]
2. Tung, W.K. *Group Theory*, 1st ed.; Cambridge University Press: New York, NY, USA, 2003.
3. Weinberg, S. *Quantum Theory of Fields Vol. 1*, 1st ed.; Cambridge University Press: New York, NY, USA, 1995.

4. Peres, A.; Scudo, P.F.; Terno, D.R. Quantum Entropy and Special Relativity. *Phys. Rev. Lett.* **2002**, *88*, 230402. [CrossRef]
5. Gingrich, R.M.; Adami, C. Quantum Entanglement of Moving Bodies. *Phys. Rev. Lett.* **2002**, *89*, 270402. [CrossRef]
6. Jordan, T.F.; Shaji, A.; Sudarshan, E.C.G. Lorentz transformations that entangle spins and entangle momenta. *Phys. Rev. A* **2007**, *75*, 022101. [CrossRef]
7. Friis, N.; Bertlmann, R.A.; Huber, M.; Hiesmayr, B.C. Relativistic entanglement of two massive particles. *Phys. Rev. A* **2010**, *81*, 042114. [CrossRef]
8. Jozsa, R.; Abrams, D.S.; Dowling, J.P.; Williams, C.P. Quantum Clock Synchronization Based on Shared Prior Entanglement. *Phys. Rev. Lett.* **2000**, *85*, 2010. [CrossRef]
9. Giovannetti, V.; Lloyd, S.; Maccone, L. Quantum-enhanced positioning and clock synchronization. *Nature* **2001**, *412*, 417–419. [CrossRef] [PubMed]
10. Kent, A.; Mundro, W.J.; Spiller, P. Quantum tagging: Authenticating location via quantum information and relativistic signaling constraints. *Phys. Rev. A* **2011**, *84*, 012326. [CrossRef]
11. Mizrahi, S.S. From qubits and actions to the Pauli-Schrödinger equation. *Phys. Scr.* **2009**, *T135*, 014007. [CrossRef]
12. Bernardini, A.E.; Mizrahi, S.S. Relativistic dynamics compels a thermalized fermi gas to a unique intrinsic parity eigenstate. *Phys. Scr.* **2014**, *89*, 075105. [CrossRef]
13. Bittencourt, V.A.S.V.; Bernardini, A.E.; Blasone, M. Global Dirac bispinor entanglement under Lorentz boosts. *Phys. Rev. A* **2018**, *97*, 032106. [CrossRef]
14. Bittencourt, V.A.S.V.; Bernardini, A.E. Entanglement of Dirac bi-spinor states driven by Poincaré classes of $SU(2) \otimes SU(2)$ coupling potentials. *Ann. Phys. (N. Y.)* **2016**, *364*, 182. [CrossRef]
15. Bittencourt, V.A.S.V.; Bernardini, A.E. Lattice-layer entanglement in Bernal-stacked bilayer graphene. *Phys. Rev. B* **2017**, *95*, 195145. [CrossRef]
16. Pal, P.B. Dirac, Majorana, and Weyl fermions. *Am. J. Phys.* **2011**, *79*, 485. [CrossRef]
17. De Leo, S.; Rotelli, P. Neutrino Chiral Oscillations. *Int. J. Theor. Phys.* **1998**, *37*, 2193. [CrossRef]
18. Bernardini, A.E. Chiral oscillations in terms of the zitterbewegung effect. *Eur. Phys. J. C* **2007**, *50*, 673–678. [CrossRef]
19. Bernardini, A.E.; De Leo, S. Flavor and chiral oscillations with Dirac wave packets. *Phys. Rev. D* **2005**, *71*, 076008. [CrossRef]
20. Nishi, C.C. First quantized approaches to neutrino oscillations and second quantization. *Phys. Rev. D* **2006**, *73*, 053013. [CrossRef]
21. Bernardini, A.E.; Guzzo, M.M.; Nishi, C. Quantum flavor oscillations extended to the Dirac theory. *Fortschr. Phys.* **2011**, *59*, 372. [CrossRef]
22. Cheng, T.-P.; Li, L.-F. *Gauge Theory of Elementary Particle Physics*, 1st ed.; Oxford Science Publication: Oxford, UK, 1995.
23. Bittencourt, V.A.S.V.; Bernardini, A.E.; Blasone, M. Chiral oscillations in the non-relativistic regime. *Eur. Phys. J. C* **2021**, *81*, 411. [CrossRef]
24. Long, A.J.; Lunardini, C.; Sabancilar, E. Detecting non-relativistic cosmic neutrinos by capture on tritium: Phenomenology and physics potential. *J. Cosmol. Astropart. Phys.* **2014**, *1408*, 038. [CrossRef]
25. Roulet, E.; Vissani, F. On the capture rates of big bang neutrinos by nuclei within the Dirac and Majorana hypotheses. *J. Cosmol. Astropart. Phys.* **2018**, *1810*, 49. [CrossRef]
26. Ge, S.-F.; Pasquini, P. Parity violation and chiral oscillation of cosmological relic neutrinos. *Phys. Lett. B* **2020**, *811*, 135961. [CrossRef]
27. Horodecki, R.; Horodecki, P.; Horodecki, M.; Horodecki, K. Quantum Entanglement. *Rev. Mod. Phys.* **2009**, *81*, 865. [CrossRef]
28. Brunner, N.; Cavalcanti, D.; Pironio, S.; Scarani, V.; Whener, S. Bell nonlocality. *Rev. Mod. Phys.* **2014**, *86*, 419. [CrossRef]
29. Thaller, B. *The Dirac Equation*, 1st ed.; Springer: Berlin, Germany, 1992.
30. da Motta, H. Chirality and neutrinos, a student first approach. *J. Phys. Conf. Ser.* **2020**, *1558*, 012014. [CrossRef]
31. Adam, A.S.; Benoit, N.J.; Kawamura, Y.; Matsuo, Y.; Morozumi, T.; Shimizu, Y.; Tokunaga, Y.; Toyota, N. Time evolution of lepton number carried by Majorana neutrinos. *arXiv* **2021**, arXiv:2101.07751.
32. Pal, P.B. *An Introductory Course of Particle Physics*, 1st ed.; CRC Press: New York, NY, USA, 2015.
33. Peres, A. Separability criterion for density matrices. *Phys. Rev. Lett.* **1996**, *77*, 1413–1415. [CrossRef]
34. Vidal, G.; Wenerm, R.F. Computable measure of entanglement. *Phys. Rev. A* **2002**, *65*, 032314. [CrossRef]
35. Greiner, W. *Relativistic Quantum Mechanics: Wave Equations*, 3rd ed.; Springer: Berlin, Germany, 2000.
36. Breuer, H.-P.; Petruccionem, F. *The Theory of Open Quantum Systems*, 1st ed.; Oxford University Press: Oxford, UK, 2002.
37. Blasone, M.; Dell'Anno, F.; De Siena, S.; Di Mauro, M.; Illuminati, F. Multipartite entangled states in particle mixing. *Phys. Rev. D* **2008**, *77*, 096002. [CrossRef]
38. Blasone, M.; Dell'Anno, F.; De Siena, S.; Illuminati, F. Entanglement in neutrino oscillations. *Europhys. Lett.* **2009**, *85*, 50002. [CrossRef]
39. Blasone, M.; Vitiello, G. Quantum field theory of fermion mixing. *Ann. Phys.* **1995**, *244*, 238–311. [CrossRef]
40. Blasone, M.; Henning, P.A.; Vitiello, G. The exact formula for neutrino oscillations. *Phys. Lett. B* **1999**, *451*, 140–145. [CrossRef]
41. Bramon, A.; Garbarino, G. Novel Bell's inequalities for entangled $K^0\bar{K}^0$ pairs. *Phys. Rev. Lett.* **2003**, *88*, 040403. [CrossRef]
42. Shi, Y. Entanglement in relativistic quantum field theory. *Phys. Rev. D* **2004**, *70*, 105001. [CrossRef]
43. Fan, J.; Li, X. Relativistic effect of entanglement in fermion-fermion scattering. *Phys. Rev. D* **2018**, *97*, 016011. [CrossRef]
44. Tanabashi, M.; Hagiwara, K.; Hikasa, K.; Nakamura, K.; Sumino, Y.; Takahashi, F.; Tanaka, J.; Agashe, K.; Aielli, G.; Amser, C.; et al. (Particle Data Group) Review of Particle Physics. *Phys. Rev. D* **2018**, *98*, 030001. [CrossRef]

45. Afik, Y.; de Nova, J.R.M. Quantum information and entanglement with top quarks at LHC. *arXiv* **2020**, arXiv:2003.02280.
46. Blasone, M.; De Siena, S.; Matrella, C. Wave packet approach to quantum correlations in neutrino oscillations. *Eur. Phys. J. C* **2021**, *81*, 660.

universe MDPI

Communication

On Gauge Invariance of the Bosonic Measure in Chiral Gauge Theories

Gabriel de Lima e Silva [1,†], Thalis José Girardi [2,†] and Sebastião Alves Dias [2,†,*]

1 Centro de Estudos Superiores de Tefé, Universidade do Estado do Amazonas, Estrada do Bexiga, 1085, Jerusalém, Tefé 69550-000, AM, Brazil; galima@uea.edu.br
2 Centro Brasileiro de Pesquisas Físicas, Rua Dr. Xavier Sigaud 150, Rio de Janeiro 22290-180, RJ, Brazil; thalisjg@gmail.com
* Correspondence: tiao@cbpf.br or sebastiao.alves.dias@gmail.com
† These authors contributed equally to this work.

Abstract: Gauge invariance of the measure associated with the gauge field is usually taken for granted, in a general gauge theory. We furnish a proof of this invariance, within Fujikawa's approach. To stress the importance of this fact, we briefly review gauge anomaly cancellation as a consequence of gauge invariance of the bosonic measure and compare this cancellation to usual results from algebraic renormalization, showing that there are no potential inconsistencies. Then, using a path integral argument, we show that a possible Jacobian for the gauge transformation has to be the identity operator, in the physical Hilbert space. We extend the argument to the complete Hilbert space by a direct calculation.

Keywords: gauge field theories; gauge anomalies; nonperturbative techniques

PACS: 11.15.-q; 11.15.Tk ; 11.30.-j

Citation: de Lima e Silva, G.; Girardi, T.J.; Dias, S.A. On Gauge Invariance of the Bosonic Measure in Chiral Gauge Theories. *Universe* **2021**, *7*, 283. https://doi.org/10.3390/universe7080283

Academic Editor: Julio Marny Hoff da Silva

Received: 31 May 2021
Accepted: 31 July 2021
Published: 4 Ausgut 2021

1. Introduction

As we celebrate the 80th anniversary of Wigner's fundamental paper [1], on how to build Poincaré symmetry in the quantum domain, it seemed relevant to us to consider some aspects of the phenomenon of quantum obstruction to implementation of a given symmetry, which is called an *anomaly*. An anomaly usually manifests as an operator that prevents the expectation value of a Noether current from vanishing. Anomalies become specially critical when they refer to gauge symmetry, usually required to prove renormalization of the corresponding gauge theories. A common statement is that gauge anomalies break Ward-Takahashi (or Slavnov-Taylor) identities, necessary to reduce the number of renormalization constants to be computed, spoiling perturbative renormalization of these theories. In this way, the unavoidable presence of a gauge anomaly is usually taken as an indication that the gauge theory under analysis is not adequate.

Chiral gauge theories are examples of this kind of theory. They are defined through minimal couplings between chiral fermions and gauge fields. As they are the basis over which the standard model is built, the solution was to choose fermion representations in such a way that gauge anomalies are canceled [2]. However, when chiral gauge theories are considered as effective theories, their gauge anomalies are not always problematic. In several contexts [3–5], with the gauge field taken as external (not quantum), gauge anomalies have been used successfully to cancel unwanted boundary contributions.

The scene described above should be enough to raise a question mark on the alleged inconsistency of chiral gauge theories. There has been a lot of work [6–9] during the 80's that showed that anomalous gauge theories are not necessarily inconsistent. A more recent work [10] indicated that, in the full quantum context (i.e., integrating also over the gauge fields), the gauge anomaly had null vacuum expectation value, in an arbitrary number of dimensions, for abelian and non-abelian theories. Its insertion in correlators of gauge invariant operators also gives a null result. These results point towards gauge symmetry restoration when the complete quantum theory is considered non-perturbatively. There are also modifications in the Ward-Takahashi identities [11], in the abelian case, which, however, do not prevent potential relations between renormalization constants, thus opening the path to prove renormalizability also in this extended context. Although this renormalizability has not yet been proved, it is clear that there is a lot of open questions concerning gauge anomalies.

This paper intends to discuss a point frequently overlooked in the literature, which concerns the gauge invariance of the *bosonic measure* in the generating functional of chiral gauge theories. In general, this is taken for granted, but no systematic analysis is found in the literature. We consider this question by using path integral arguments to show that it is indeed gauge invariant. We check our findings with known results from the renormalization of Yang-Mills theories. Given the central role played by this argument in several fundamental instances (e.g., Faddeev-Popov's method), we believe that this study may fill an important gap in the area. We organize our discussion as follows: in Section 2, we briefly review the gauge anomaly vanishing mechanism, for the general (non-abelian) case. We also consider arguments against this vanishing, from an algebraic renormalization approach, and show that they do not apply to this context. The role of gauge invariance of the bosonic measure is stressed. In Section 3, we derive, by a path integral argument, the gauge invariance of the bosonic measure, when we restrict ourselves to the physical Hilbert space. We extend our argument to the whole Hilbert space by performing a calculation based on Fujikawa's approach to Jacobians. In Section 4, we present our conclusions and some future perspectives.

2. Is There a Gauge Anomaly in Chiral Gauge Theories?

In order to fix our conventions and define precisely the problem under investigation, we briefly review some of the main results of reference [10], pointing the role played by the invariance of the bosonic measure when appropriate. What we call *chiral gauge theories* are described by an action $S[\psi, \bar{\psi}, A_\mu]$, given by

$$\begin{aligned} S[\psi, \bar{\psi}, A_\mu] &= S_G[A_\mu] + S_F[\psi, \bar{\psi}, A_\mu] \\ &= \int dx \, \frac{1}{2} \mathrm{tr}\, F_{\mu\nu} F^{\mu\nu} + \int dx \, \bar{\psi} D \psi, \end{aligned} \tag{1}$$

where dx indicates integration over a d-dimensional Minkowski space. The operator D is called the Dirac operator of the theory and is given by

$$D = i\gamma^\mu (\partial_\mu \mathbf{1} - ieA_\mu) \equiv i\gamma^\mu D_\mu. \tag{2}$$

The fields ψ are *left handed Weyl fermions* ($\gamma_5 \psi = \psi$) carrying the fundamental representation of $SU(N)$. As usual, A_μ takes values in the Lie algebra of $SU(N)$ such that

$$\begin{aligned} A_\mu &= A_\mu^a T_a, \\ F_{\mu\nu} &= \partial_\mu A_\nu - \partial_\nu A_\mu - ie[A_\mu, A_\nu], \end{aligned} \tag{3}$$

and the generators T_a satisfy

$$[T_a, T_b] = i f_{abc} T_c, \qquad \mathrm{tr}\,(T_a T_b) = -\frac{1}{2}\delta_{ab}. \tag{4}$$

Considering

$$g = \exp(i\theta^a(x) T_a), \tag{5}$$

and simultaneous changes of the fields ψ and A_μ as

$$
\begin{aligned}
A_\mu^g &= gA_\mu g^{-1} + \frac{i}{e}(\partial_\mu g)g^{-1}, \\
\psi^g &= g\psi, \\
\bar{\psi}^g &= \bar{\psi}g^{-1},
\end{aligned} \tag{6}
$$

the action S is classically gauge invariant

$$
S[\psi^g, \bar{\psi}^g, A_\mu^g] = S[\psi, \bar{\psi}, A_\mu]. \tag{7}
$$

2.1. *The Gauge Anomaly and Its Vanishing*

The quantum theory is defined by the generating functional, which is

$$
Z[\eta, \bar{\eta}, j_a^\mu] = \int d\psi d\bar{\psi} dA_\mu \exp\left(iS[\psi, \bar{\psi}, A_\mu] + i \int dx [\bar{\eta}\psi + \bar{\psi}\eta + j_a^\mu A_\mu^a] \right). \tag{8}
$$

We set the external sources to zero and perform a change of variables $A_\mu \longrightarrow A_\mu^g$:

$$
\begin{aligned}
Z[0,0,0] &= \int d\psi d\bar{\psi} dA_\mu \exp(iS[\psi, \bar{\psi}, A_\mu]) \\
&= \int d\psi d\bar{\psi} dA_\mu^g \exp\left(iS[\psi, \bar{\psi}, A_\mu^g]\right).
\end{aligned} \tag{9}
$$

It is well known [12] that the fermion measure is not invariant under $\psi \longrightarrow \psi^g$, $\bar{\psi} \longrightarrow \bar{\psi}^g$. In what concerns the bosonic measure, its gauge invariance is universally aceppted

$$
dA_\mu^g = dA_\mu. \tag{10}
$$

We show, below, that once Equation (10) is considered, we can promptly show that the Noether current associated to the gauge symmetry (6) is covariantly conserved. To see this, suppose $g = 1 + i\theta^a(x)T_a$, with θ^a infinitesimal, and remember that

$$
A_\mu^g = A_\mu - \frac{1}{e}\mathcal{D}_\mu\theta, \tag{11}
$$

with

$$
\mathcal{D}_\mu\theta = T_a\left(\partial_\mu\delta_b^a + ef_{abc}A_\mu^c\right)\theta^b \equiv T_a(\mathcal{D}_\mu)^a{}_b\theta^b. \tag{12}
$$

Using this expression for A_μ^g (10) we obtain:

$$
\begin{aligned}
Z[0,0,0] &= \int d\psi d\bar{\psi} dA_\mu^g \exp\left(iS[\psi, \bar{\psi}, A_\mu^g]\right) \\
&= \int d\psi d\bar{\psi} dA_\mu \exp\left(iS[\psi, \bar{\psi}, A_\mu] + \int dx\, \theta^a(x)(\mathcal{D}_\mu)^a{}_b(\bar{\psi}\gamma^\mu T_b\psi)\right) \\
&= Z[0,0,0] \\
&+ \int dx\, \theta^a(x) \int d\psi d\bar{\psi} dA_\mu \left[(\mathcal{D}_\mu)^a{}_b(\bar{\psi}\gamma^\mu T_b\psi)\right] \exp(iS[\psi, \bar{\psi}, A_\mu]).
\end{aligned} \tag{13}
$$

This means

$$
\begin{aligned}
&\int d\psi d\bar{\psi} dA_\mu \left[(\mathcal{D}_\mu)^a{}_b(\bar{\psi}\gamma^\mu T_b\psi)\right] \exp(iS[\psi, \bar{\psi}, A_\mu]) \\
&= \langle 0|(\mathcal{D}_\mu)^a{}_b(\bar{\psi}\gamma^\mu T_b\psi)|0\rangle \\
&= 0.
\end{aligned} \tag{14}
$$

We could follow a different path when considering the original dependence of Z on A^g_μ, by absorbing the gauge dependence in the fermions and *using again the bosonic measure gauge invariance*:

$$\begin{aligned} Z[0,0,0] &= \int d\psi d\bar{\psi} dA^g_\mu \exp\left(iS[\psi,\bar{\psi},A^g_\mu]\right) \\ &= \int d\psi d\bar{\psi} dA_\mu \exp\left(iS[\psi^{g^{-1}},\bar{\psi}^{g^{-1}},A_\mu]\right) \\ &= \int d\psi d\bar{\psi} dA_\mu \exp\left(iS[\psi,\bar{\psi},A_\mu] - i\alpha_1\left[A_\mu,g^{-1}\right]\right), \end{aligned} \tag{15}$$

where $\alpha_1\left[A_\mu,g^{-1}\right]$ is related to the Jacobian for the gauge transformation of the fermionic measure as,

$$d\psi d\bar{\psi} = \exp(-i\alpha_1\left[A_\mu,g^{-1}\right])d\psi^{g^{-1}}d\bar{\psi}^{g^{-1}}. \tag{16}$$

Under an infinitesimal gauge transformation,

$$\alpha_1(A_\mu,-\theta) = i\int dx\,\theta^a\mathcal{A}^a(A_\mu) + \ldots, \tag{17}$$

with $\mathcal{A}_a(A_\mu)$ being the gauge anomaly operator, as is well known [12]. Following this path we arrive at

$$\begin{aligned} &Z[0,0,0] = Z[0,0,0] \\ &- i\int dx\,\theta^a \int d\psi d\bar{\psi} dA_\mu \mathcal{A}^a(A_\mu)\exp(iS[\psi,\bar{\psi},A_\mu]) \\ &\Rightarrow \int d\psi d\bar{\psi} dA_\mu \mathcal{A}_a(A_\mu)\exp(iS[\psi,\bar{\psi},A_\mu]) \\ &= \langle 0|\mathcal{A}^a(A_\mu)|0\rangle = 0. \end{aligned} \tag{18}$$

If the integration over the gauge fields were not performed, it is easy to see that the result would be

$$\langle 0|(\mathcal{D}_\mu)^a{}_b(\bar{\psi}\gamma^\mu T_b\psi)|0\rangle_{A_\mu} = \mathcal{A}^a(A_\mu), \tag{19}$$

where

$$\langle 0|(\mathcal{D}_\mu)^a{}_b(\bar{\psi}\gamma^\mu T_b\psi)|0\rangle_{A_\mu} = \int d\psi d\bar{\psi}\left[(\mathcal{D}_\mu)^a{}_b(\bar{\psi}\gamma^\mu T_b\psi)\right]\exp(iS[\psi,\bar{\psi},A_\mu]), \tag{20}$$

corresponds to the situation of the chiral fermions being considered under the influence of a fixed external field A_μ.

This means that, *with all the fields being quantized*, there is no gauge anomaly preventing the conservation of the gauge current $J^\mu_a = \bar{\psi}\gamma^\mu T_b\psi$. We have to stress that this result is not in contradiction with the well-known existence and topological interpretation of the gauge anomaly (see, for example, [13,14]), since it is always present when the integration over the fields A^a_μ is not performed (i.e., they are taken as external fields). However, when quantum corrections are taken into account, the simple argument above shows that it must vanish.

If the bosonic measure dA_μ were not gauge invariant, we should have to add the contribution of the Jacobian of the gauge transformation of that measure, which would spoil the covariant conservation just obtained. We will further investigate this gauge invariance in the next section. Note also that the absence of gauge invariance of the fermionic measure plays no role in the covariant conservation of the fully quantized gauge current, as opposed to usual statements. The vanishing of the vacuum expectation value of the anomaly is to be seen as a consistency requirement, in the situation of a fully quantized gauge field.

2.2. An Algebraic Renormalization Objection

An argument, based on the technique of algebraic renormalization [15], could in principle question the findings reported above. One starts with a gauge fixed version of (1), namely

$$S_{\text{FP}}[\psi,\bar{\psi},A_\mu,B,c,\bar{c}] = S[\psi,\bar{\psi},A_\mu] + S_{\text{gf}}[A_\mu,B,c,\bar{c}], \tag{21}$$

where

$$S_{\text{gf}}[A_\mu,B,c,\bar{c}] = \int dx\, \text{tr}\left(B\partial^\mu A_\mu + \frac{1}{2}\alpha B^2 - \bar{c}\partial^\mu\left(i\partial_\mu + ie[c,A_\mu]\right)\right), \tag{22}$$

with B being an auxiliary field (the Lautrup-Nakanishi field) and c, $\bar{c}$ being ghost fields, all of them taking values on the Lie algebra of $SU(N)$. The action (21) would be obtained from (1) by the use of the Faddeev-Popov procedure to fix the gauge. Then, one defines the linearized Slavnov-Taylor operator [16], which will test gauge invariance of the (quantum) effective action at each order of perturbation theory. When one solves the cohomology problem associated with this nilpotent operator, one finds a non-trivial solution. This means that the effective action is found to be not gauge invariant, as gauge symmetry violating terms dynamically generated at any given order can not be absorbed by suitably chosen gauge invariant counterterms. So, gauge invariance would be hopelessly lost and a gauge anomaly would be present, contradicting frontally what we found previously and putting renormalizability at risk.

There is nothing wrong with this line of reasoning, except its starting point. Gauge fixing is essential for the perturbative definition of a gauge theory, but one can not do it in a gauge anomalous theory in the same way as one does in the case of a non gauge anomalous one. If one starts with the non gauge fixed action (1) and insists in the insertion of the identity through the Faddeev-Popov's method, one ends up with a generating functional given by [8]

$$\begin{aligned} Z[0,0,0] &= \int d\theta d\psi d\bar{\psi} dA_\mu dB dc d\bar{c} \\ &\quad \times \exp\left(iS_{\text{FP}}[\psi,\bar{\psi},A_\mu,B,c,\bar{c}] + i\alpha_1(A_\mu,\theta)\right) \\ &= \int d\theta d\psi d\bar{\psi} dA_\mu dB dc d\bar{c} \exp\left(iS_{\text{full}}[\psi,\bar{\psi},A_\mu,B,c,\bar{c},\theta]\right), \end{aligned} \tag{23}$$

where α_1 was defined in (16) and is called *Wess-Zumino action*. The fields $\theta = \theta^a T_a$ are called Wess-Zumino fields and represent new quantum degrees of freedom. The action S_{full} is the true starting point from where one should restart the analysis of the cohomology of the linearized Slavnov-Taylor operator. It is gauge invariant and, thanks to Faddeev-Popov's technique, it has a well-defined gauge boson propagator. Unfortunately, as the fields θ^a have null mass dimension, it is not known how to perform this analysis up to now. In order to avoid the appearance of the Wess-Zumino fields, one is not allowed to fix the gauge. However, in doing this, there is no BRS symmetry to help one with the analysis of gauge invariance at an arbitrary perturbative order.

Thus, within the present knowledge, algebraic renormalization methods do not seem to be useful to decide if chiral gauge theories are truly gauge anomalous (and potentially inconsistent). As we pointed in the previous section, there are strong indications of the opposite.

3. On the Gauge Invariance of the Bosonic Measure

Let us now focus on the behavior of the bosonic measure under gauge transformations. To this end, let us first display a preparatory argument: consider the generating functional for correlators of gauge invariant operators in pure Yang-Mills theory (without chiral fermions). These gauge invariant operators satisfy

$$O_i\left(A_\mu^g\right) = O_i(A_\mu) \tag{24}$$

and the correlators are obtained as

$$\frac{\delta^n}{\delta\lambda^1(x_1)\ldots\delta\lambda^n(x_n)}Z\left[\lambda^i\right]\bigg|_{\lambda^i=0} = \langle 0|T(O_1(A_\mu)(x_1)\ldots O_n(A_\mu)(x_n))|0\rangle, \tag{25}$$

with

$$Z\left[\lambda^i\right] = \int dA_\mu \exp i\int \mathrm{tr}\left(\frac{1}{2}F_{\mu\nu}F^{\mu\nu} + \lambda^i O_i[A_\mu]\right). \tag{26}$$

Considering the integration not over A_μ but over its gauge transformed version A^g_μ:

$$\begin{aligned}
Z\left[\lambda^i\right] &= \int dA_\mu \exp i\int \mathrm{tr}\left(\frac{1}{2}F_{\mu\nu}F^{\mu\nu} + \lambda^i O_i(A_\mu)\right) \\
&= \int dA^g_\mu \exp i\int \mathrm{tr}\left(\frac{1}{2}(F_{\mu\nu}F^{\mu\nu})^g + \lambda^i O_i\left(A^g_\mu\right)\right) \\
&= \int dA_\mu J[A_\mu, g]\exp\left(i\int \mathrm{tr}\left(F_{\mu\nu}F^{\mu\nu} + \lambda^i O_i(A_\mu)\right)\right),
\end{aligned} \tag{27}$$

where we allowed the potential presence of a Jacobian for the gauge transformation of the measure. What was obtained above corresponds to,

$$\begin{aligned}
&\langle 0|T(J[A_\mu, g]O_1(A_\mu)(x_1)\ldots O_n(A_\mu)(x_n))|0\rangle \\
&= \langle 0|T(O_1(A_\mu)(x_1)\ldots O_n(A_\mu)(x_n))|0\rangle,
\end{aligned} \tag{28}$$

which, translated into words, means that all correlators involving $J\left[A_\mu, g\right]$ with gauge invariant operators are the same as those involving the identity. Thus, in the physical Hilbert space of the theory, the two operators are the same.

This argument does not generalize to arbitrary, non-gauge invariant operators. However, an explicit calculation can resolve the problem. Let us use the usual prescription of defining the bosonic measure by means of a complete set of orthonormal eigenfunctions $\{\phi_n\}$ of an hermitian operator $\bar{D}$:

$$\begin{aligned}
\bar{D}\phi_n &= \lambda_n\phi_n, \\
\int dx\phi^\dagger_n\phi_m &= \delta_{nm}, \quad \sum_n \phi_n(x)\phi^\dagger_n(y) = \delta(x-y), \\
A^a_\mu(x) &= \sum_n a^a_{\mu,n}\phi_n(x) \to dA_\mu = \prod_{a,\mu,n} da^a_{\mu,n}.
\end{aligned} \tag{29}$$

Under a infinitesimal gauge transformation (11) we have

$$\begin{aligned}
A^g_\mu &= \sum_n \bar{a}^a_{\mu,n}T_a\phi_n(x) = \sum_n a^a_{\mu,n}T_a\phi_n(x) - \frac{i}{e}\mathcal{D}_\mu\theta \\
&= \left(\sum_n\left(a^a_{\mu,n} + ia^b_{\mu,n}f_{abc}\theta^c\right)\phi_n(x) - \frac{i}{e}\partial_\mu\theta^a\right)T_a.
\end{aligned} \tag{30}$$

Decomposing θ^a in terms of the same eigenfunctions of $\bar{D}$,

$$-\frac{i}{e}\partial_\mu\theta^a(x) = \sum_n \tilde{a}^a_{\mu,n}\phi_n(x), \tag{31}$$

We obtain,

$$\bar{a}^a_{\mu,n} = \sum_m\left(\delta_{ab}\delta_{nm} + \int dx\phi^\dagger_n(x)if_{abc}\theta^c(x)\phi_m(x)\right)a^b_{\mu,m} + \tilde{a}^a_{\mu,n}, \tag{32}$$

so that

$$\prod_{a,\mu,n} d\bar{a}^a_{\mu,n} = \det\left[\delta_{ab}\delta_{nm} + \int dx\phi_n^\dagger(x)if_{abc}\theta^c(x)\phi_m(x)\right]\prod_{a,\mu,n} da^a_{\mu,n}. \tag{33}$$

The term $\tilde{a}^a_{\mu,n}$ does not contribute because of translational invariance of each measure $da^a_{\mu,n}$. Following the steps of Fujikawa [12] we get the expression for the Jacobian:

$$J[A_\mu,\theta] = \exp\left(\sum_n\left(\mathrm{tr}\int dx\phi_n^\dagger(x)if_{abc}\theta^c(x)\phi_n(x)\right)\right) \tag{34}$$

where "tr" is to be computed over Lie algebra indices. It is easy to see that the expression for $J[A_\mu,\theta]$ is indefinite:

$$\begin{aligned}\sum_n\left(\mathrm{tr}\int dx\phi_n^\dagger(x)if_{abc}\theta^c(x)\phi_n(x)\right) &= \mathrm{tr}\int dx if_{abc}\theta^c(x)\sum_n\phi_n(x)\phi_n^\dagger(x)\\ &= \int dx if_{aac}\theta^c(x)\delta(0) = 0\times\infty.\end{aligned} \tag{35}$$

So, it must be regularized in order to make sense. It is natural to choose the eigenvalues of the operator $\bar{D}$ to regularize the Jacobian as

$$\begin{aligned}J[A_\mu,\theta] &\equiv \exp\left(\lim_{M^2\to\infty}\sum_n\left(\mathrm{tr}\int dx\phi_n^\dagger(x)if_{abc}\theta^c(x)\exp\left(-\frac{\lambda_n^2}{M^\alpha}\right)\phi_n(x)\right)\right)\\ &= \exp\left(\lim_{M^2\to\infty}\sum_n\left(\mathrm{tr}\int dx\phi_n^\dagger(x)if_{abc}\theta^c(x)\exp\left(-\frac{\bar{D}^2}{M^\alpha}\right)\phi_n(x)\right)\right),\end{aligned} \tag{36}$$

where α is chosen so that the argument of the exponential is dimensionless. The choice of the operator $\bar{D}$ is usually guided by the requisites that (a) it naturally appears in the theory; (b) it is gauge invariant; and (c) its eigenvalues are real. Besides, our choice of the coefficients $a^a_{\mu,n}$ carrying all the dependence on μ and a implies that the ϕ_n must be eigenfunctions of an scalar colorless operator. A good choice is

$$\bar{D} = \mathrm{tr}\,(\mathcal{D}_\mu\mathcal{D}^\mu), \tag{37}$$

where the trace is taken only over the color indices. Under these conditions, we see that the sum is regularized and, as there is no additional dependence on color indices coming from $\exp(-\bar{D}^2/M^4)$, the trace can be immediately taken and the result is

$$\begin{aligned}J[A_\mu,\theta] &= \exp\left(\lim_{M^2\to\infty}\sum_n\left(if_{aac}\int dx\phi_n^\dagger(x)\theta^c(x)\exp\left(-\frac{\bar{D}^2}{M^4}\right)\phi_n(x)\right)\right)\\ &= \exp(0) = 1.\end{aligned} \tag{38}$$

Of course, one could choose other strategies and a result different from 1 could arise. However, the "gauge anomaly" coming from this "non-trivial" Jacobian could be removed by a adequate choice of counterterms. To say this more precisely, we can use what we know from the fact that Yang-Mills theories are renormalizable. In fact, 't Hooft's proof [17,18] shows that it is possible to preserve gauge invariance at every order in perturbation theory and this is crucial for the demonstration that the theory is renormalizable. Algebraic renormalization results confirm this, by noticing that the cohomology of the Slavnov-Taylor operator is trivial for a Yang-Mills theory [19]. Then, even if we would regularize the theory with non-gauge invariant regulators (then obtaining a non-trivial Jacobian), a change in renormalization scheme could restore gauge invariance and set the Jacobian as 1.

4. Conclusions

Gauge invariance of the bosonic measure is essential to cancel the gauge anomaly at full quantum level in a chiral gauge theory (abelian or non-abelian). Besides this, it is crucial for the implementation of Fadeev-Popov's technique. Although it is a commonly used feature, this gauge invariance has not deserved a careful consideration in the literature (up to our knowledge). This paper intends to furnish a detailed analysis of this useful property and, in so doing, to fill an important omission. It inserts itself into a program of investigation of the renormalization of chiral gauge theories, to see if it can really be implemented at perturbative level. Although the vacuum expectation value of the gauge anomaly vanishes, there are indications that it has non-null insertions in correlators of non-gauge invariant operators. The dynamical picture remains unclear.

An important point would be to ask what could happen to the axial anomaly, defined as

$$\langle 0|\partial_\mu J_5^\mu|0\rangle_{A_\mu} = \langle 0|\partial_\mu(\bar{\psi}\gamma^\mu\gamma_5\psi)|0\rangle_{A_\mu} = \mathcal{A}_5(A_\mu).$$

It is related to the quantum violation of axial symmetry

$$\psi' = e^{\alpha\gamma_5}\psi \equiv g_5\psi,$$

which is classically exact in the massless case. The gauge anomaly appears as the non-covariant conservation of the gauge current

$$\langle 0|\left(D_\mu\right)^a{}_b J_b^\mu|0\rangle_{A_\mu} = \langle 0|\left(D_\mu\right)^a{}_b(\bar{\psi}\gamma^\mu T_b\psi)|0\rangle_{A_\mu} = \mathcal{A}^a(A_\mu),$$

Whose classical conservation would be a consequence of gauge symmetry

$$\psi' = e^{i\theta^a T_a}\psi \equiv g\psi. \tag{39}$$

While the gauge transformation (39) can be transferred from the fermions to the gauge fields (with all of them being considered as quantum)

$$S\left[\psi^g, \bar{\psi}^g, A_\mu\right] = S\left[\psi, \bar{\psi}, A_\mu^{g^{-1}}\right],$$

the same can not be done with the axial transformation under the same circumstances:

$$S\left[\psi^{g_5}, \bar{\psi}^{g_5}, A_\mu\right] = S\left[\psi, \bar{\psi}, A_\mu\right],$$

which means that the gauge fields do not play any role in axial symmetry. As a consequence, we are not allowed to make the same manipulations in the functional integral and so, we can not say that the axial anomaly is canceled as well. The axial anomaly is a welcome one, as it is responsible for the explanation of the observed π^0 decay into two photons. On the other hand, the gauge anomaly is a generally undesired feature and has to be canceled (in the construction of the standard model) in order not to ruin explicit renormalizability of the theory. So, we could answer that the dynamical role of the axial anomaly is unaltered in our approach. It keeps appearing and influencing low energy effective actions of QCD, for example. However, we offer some hope that the gauge anomaly is not such a catastrophic issue.

We must also remind the reader that there must be several ways to prove gauge invariance of the bosonic measure. One of them is surely to consider the theory on the lattice, where one can define the bosonic measure as a Haar measure, which is naturally gauge invariant (see, for example, Formula (7.12) of [20]). Then, we could consider the limit of zero lattice spacing with due caution to establish this invariance in the continuum theory. However, this discussion (and other ones, using different arguments) is not usually seen in the literature and we found it could be useful to present it explicitly, by means of a Fujikawa approach.

Some big issues have to be considered in this path. Perhaps one of the most urgent would be how to define the tree level of chiral gauge theories. Since one is not allowed to fix the gauge (due to the appearance of a Jacobian for the fermion measure), we have no definition of a free gauge boson propagator, which invalidates a perturbative analysis of the theory. We also have no natural choice to choose a method of regularization to deal with loops, as gauge invariance can not be used as a guiding principle. We will continue to investigate these and other issues and our findings will be reported as they appear.

Author Contributions: Conceptualization, G.d.L.e.S., T.J.G. and S.A.D.; methodology, G.d.L.e.S., T.J.G. and S.A.D.; software, G.d.L.e.S., T.J.G. and S.A.D.; validation, G.d.L.e.S., T.J.G. and S.A.D.; formal analysis, G.d.L.e.S., T.J.G. and S.A.D.; investigation, G.d.L.e.S., T.J.G. and S.A.D.; resources, G.d.L.e.S., T.J.G. and S.A.D.; data curation, G.d.L.e.S., T.J.G. and S.A.D.; writing—original draft preparation, G.d.L.e.S., T.J.G. and S.A.D.; writing—review and editing, G.d.L.e.S., T.J.G. and S.A.D.; visualization, G.d.L.e.S., T.J.G. and S.A.D.; supervision, G.d.L.e.S., T.J.G. and S.A.D.; project administration, G.d.L.e.S., T.J.G. and S.A.D.; funding acquisition, G.d.L.e.S., T.J.G. and S.A.D. All authors have read and agreed to the published version of the manuscript.

Funding: T. J. Girardi was funded by Coordenação de Aperfeiçoamento de Pessoal de Nível Superior—CAPES, Brazil. G. de Lima e Silva was funded by Fundação de Amparo à Pesquisa do Estado do Amazonas—FAPEAM, Brazil.

Institutional Review Board Statement: Not applicable.

Informed Consent Statement: Not applicable.

Data Availability Statement: Not applicable.

Conflicts of Interest: The authors declare no conflict of interest. The funders had no role in the design of the study; in the collection, analyses, or interpretation of data; in the writing of the manuscript, or in the decision to publish the results.

References

1. Wigner, E. On unitary representations of the inhomogeneous Lorentz group. *Ann. Math.* **1939**, *40*, 149–204. [CrossRef]
2. Lüsher, M. Chiral gauge theories revisited, *arXiv* **2001**, arXiv:0102028.
3. Wen, X.G. Gapless boundary excitations in the quantum Hall states and in the chiral spin states. *Phys. Rev. B* **1991**, *43*, 11025–11036. [CrossRef] [PubMed]
4. Robinson, S.P.; Wilczek, F. Relationship between Hawking radiation and gravitational anomalies. *Phys. Rev. Lett.* **2005**, *95*, 011303. [CrossRef] [PubMed]
5. Lima, G.L.S.; Dias, S.A. Edge modes in the fractional quantum Hall effect without extra edge fermions. *Europhys. Lett.* **2011**, *94*, 37005. [CrossRef]
6. Jackiw, R.; Rajaraman, R. Vector-meson mass generation by chiral anomalies. *Phys. Rev. Lett.* **1985**, *54*, 1219–1221. [CrossRef] [PubMed]
7. Faddeev, L.D.; Shatashvili, S.L. Realization of the Schwinger term in the Gauss law and the possibility of correct quantization of a theory with anomalies. *Phys. Lett. B* **1986**, *167*, 225–228. [CrossRef]
8. Harada, K.; Tsutsui, I. On the path-integral quantization of anomalous gauge theories. *Phys. Lett. B*, **1987**, *183*, 311314. [CrossRef]
9. Babelon, O.; Shaposnik, F.; Viallet, C. Quantization of gauge theories with Weyl fermions. *Phys. Lett. B*, **1986**, *177*, 385–388. [CrossRef]
10. Santiago Lima, G.L.; Chaves, R.; Dias, S.A. Gauge anomaly cancellation in chiral gauge theories. *Ann. Phys.* **2012**, *327*, 1435–1449. [CrossRef]
11. de Lima, A.P.C.R.; Dias, S.A. Ward–Takahashi identities for abelian chiral gauge theories. *Int. J. Mod. Phys. A* **2016**, *31*, 1650062. [CrossRef]
12. Fujikawa, K.; Suzuki, H. *Path Integrals and Quantum Anomalies*, 1st ed.; The International Series of Monographs on Physics 122; Oxford University Press: New York, NY, USA, 2004; pp. 65–91.
13. Alvarez-Gaumé, L.; Witten, E. Gravitational anomalies. *Nucl. Phys. B* **1984**, *234*, 269–330. [CrossRef]
14. Alvarez-Gaumé, L.; Ginsparg, P. The structure of gauge and gravitational anomalies. *Ann. Phys.* **1985**, *161*, 423–490. [CrossRef]
15. Piguet, O.; Sorella, S. *Algebraic Renormalization*, 1st ed.; Lecture Notes in Physics—Monographs M28; Springer: Berlin/Heidelberg, Germany, 1995; pp. 21–80.
16. Piguet, O.; Rouet, A. Symmetries in perturbative quantum field theory. *Phys. Rep.* **1981**, *76*, 1–77. [CrossRef]
17. 't Hooft, G. Renormalization of massless Yang-Mills fields. *Nucl. Phys. B* **1971**, *33*, 173–199.
18. 't Hooft, G. Renormalizable Lagrangians for massive Yang-Mills fields. *Nucl. Phys. B* **1971**, *35*, 167–188. [CrossRef]

19. Becchi, C.; Rouet, A.; Stora, R. Renormalization of gauge theories. *Ann. Phys.* **1976**, *98*, 287–321. [CrossRef]
20. Creutz, M. *Quarks, Gluons and Lattices*; Cambridge University Press: New York, NY, USA, 1983.

universe

MDPI

Article

Relativistic Combination of Non-Collinear 3-Velocities Using Quaternions

Thomas Berry *,† **and Matt Visser** †

School of Mathematics and Statistics, Victoria University of Wellington, PO Box 600, Wellington 6140, New Zealand; matt.visser@sms.vuw.ac.nz

* Correspondence: thomas.berry@sms.vuw.ac.nz

† These authors contributed equally to this work.

Received: 18 November 2020; Accepted: 8 December 2020; Published: 11 December 2020

Abstract: Quaternions have an (over a century-old) extensive and quite complicated interaction with special relativity. Since quaternions are intrinsically 4-dimensional, and do such a good job of handling 3-dimensional rotations, the hope has always been that the use of quaternions would simplify some of the algebra of the Lorentz transformations. Herein we report a new and relatively nice result for the relativistic combination of non-collinear 3-velocities. We work with the relativistic half-velocities w defined by $v = \frac{2w}{1+w^2}$, so that $w = \frac{v}{1+\sqrt{1-v^2}} = \frac{v}{2} + \mathcal{O}(v^3)$, and promote them to quaternions using $\mathbf{w} = w\ \hat{\mathbf{n}}$, where $\hat{\mathbf{n}}$ is a unit quaternion. We shall first show that the composition of relativistic half-velocities is given by $\mathbf{w}_{1\oplus 2} \equiv \mathbf{w}_1 \oplus \mathbf{w}_2 \equiv (1-\mathbf{w}_1\mathbf{w}_2)^{-1}(\mathbf{w}_1+\mathbf{w}_2)$, and then show that this is also equivalent to $\mathbf{w}_{1\oplus 2} = (\mathbf{w}_1+\mathbf{w}_2)(1-\mathbf{w}_2\mathbf{w}_1)^{-1}$. Here as usual we adopt units where the speed of light is set to unity. Note that all of the complicated angular dependence for relativistic combination of non-collinear 3-velocities is now encoded in the quaternion multiplication of $\mathbf{w}_1$ with $\mathbf{w}_2$. This result can furthermore be extended to obtain novel elegant and compact formulae for both the associated Wigner angle Ω and the direction of the combined velocities: $e^{\mathbf{\Omega}} = e^{\Omega\,\hat{\mathbf{\Omega}}} = (1-\mathbf{w}_1\mathbf{w}_2)^{-1}(1-\mathbf{w}_2\mathbf{w}_1)$, and $\hat{\mathbf{w}}_{1\oplus 2} = e^{\mathbf{\Omega}/2}\,\frac{\mathbf{w}_1+\mathbf{w}_2}{|\mathbf{w}_1+\mathbf{w}_2|}$. Finally, we use this formalism to investigate the conditions under which the relativistic composition of 3-velocities is associative. Thus, we would argue, many key results that are ultimately due to the non-commutativity of non-collinear boosts can be easily rephrased in terms of the non-commutative algebra of quaternions.

Keywords: special relativity; combination of velocities; wigner angle; quaternions

1. Introduction

Hamilton first described the quaternions in the mid-1800s, primarily with a view to finding algebraically simple ways to handle 3-dimensional rotations. With the advent of special relativity in 1905, and noting the manifestly 4-dimensional nature of quaternions once one adds a real part, multiple authors have tried to interpret special relativity in an intrinsically quaternionic fashion [1–9].

Despite technical success in applying quaternions to special relativity, the use of quaternions in this subject has never really gained all that much traction in the physics community. Perhaps one of the reasons for this is that there are a number of sub-optimal notational choices in Silberstein's original work [1–3], and the fact that there is no generally accepted way of using quaternions to represent Lorentz transformations, with many different authors employing their own quite distinct methods [1–9]. Even in more recent, post-millennial, articles on "quaternionic special relativity" there is considerable disagreement on notational choices [10–13].

Below we shall introduce what we feel is a particularly simple and straightforward method for combining relativistic 3-velocities using quaternions. In particular, we shall present some new and compact formulae for computing the Wigner angle [14]. All of the interesting features due to non-commutativity properties of non-collinear boosts are implicitly and rather efficiently dealt with by the non-commutative algebra of quaternions. The method is based on an extension of an analysis by Giust, Vigoureux, and Lages [15,16], who (because they were working with the usual complex numbers) were essentially limited to motion in 2-space; their formalism is not really well-adapted to general motions in 3-space. Related constructions can also be found in [10,11].

Observe that there is a representation of pure quaternions in terms of a subset of 2×2 matrices, specifically the anti-hermitian 2×2 matrices, essentially $\sqrt{-1} \times$ (Pauli matrices). (The factor of $\sqrt{-1}$ is important.) However this does not mean that replacing quaternions by Pauli matrices in any way simplifies our results below; it just complicates the formalism. Neither does this mean that any of our results below are at all "well-known" in this alternate notation. We have carefully checked the relevant literature, (roughly speaking, 2-spinor representations of the Lorentz group). There is much more than pedagogy going on—the results reported in our article are (apart from a consistency check or two) both novel and interesting (see also [13].)

2. Preliminaries

2.1. Lorentz Transformations

The set of all Lorentz transformations of space-time form a group called the Lorentz group. Mathematically, the Lorentz group is isomorphic to $O(1,3)$, the orthogonal group of one time and three space dimensions that preserves the space-time interval

$$s^2 = -t^2 + x^2 + y^2 + z^2. \tag{1}$$

Here and hereafter, as usual we adopt units where the speed of light is set to unity. It is clear from this description that rotations of space-time are included in the Lorentz group, as well as the more familiar pure Lorentz transformations (boosts). In fact, the pure Lorentz transformations do not even form a subgroup of the Lorentz group as, in general, the composition of two boosts B_1 and B_2 is not another boost but in fact a boost and a rotation $B_{12}R_{12} = B_1B_2$; while $B_{21}R_{21} = B_2B_1$. This rotation, known as the Wigner rotation, was first discovered by Llewellyn Thomas in 1926 whilst trying to describe the Zeeman effect from a relativistic view-point [17], and was more fully analyzed by Eugene Wigner in 1939 [14]. (For more recent discussions see [18–22]).

It is well–known that the composition of Lorentz transformations is non-commutative. That is, applying two successive boosts B_1 and B_2 in different orders results in the same final boost, $B_{12} = B_{21}$, but different rotations, $R_{12} \neq R_{21}$. In the context of the combination of two velocities $\vec{v}_1$ and $\vec{v}_2$, this means that the final speed is the same no matter the order we combine the velocities, $||\vec{v}_1 \oplus \vec{v}_2|| = ||\vec{v}_2 \oplus \vec{v}_1||$, but the final directions they point in are different $\hat{v}_{1\oplus 2} \neq \hat{v}_{2\oplus 1}$. Although not immediately obvious, the angle between $\vec{v}_{1\oplus 2} = \vec{v}_1 \oplus \vec{v}_2$ and $\vec{v}_{2\oplus 1} = \vec{v}_2 \oplus \vec{v}_1$ is in fact the Wigner angle Ω, see Reference [22]. The Lorentz group has very many different representations, one of which is formulated by using the quaternions [1,2,4].

One could instead try to deal with the non-commutativity of the Lorentz transformations by adapting the general formalism of the Baker–Campbell–Hausdorff theorem [23–27]. Unfortunately the general BCH formalism applied to this problem very quickly becomes intractable, and we have found that the specifics of the quaternion formalism yield much more useful and tractable results.

Since the full symmetry group of the Maxwell equations is the conformal extension of the Poincare group, it is sometimes useful, (when looking at pure electromagnetic effects), to work with this conformal extension. However physical observers, (physical clocks and physical rulers), break the conformal invariance, and to even meaningfully define 3-velocities one needs to restrict attention to the Poincare group. We shall go even further and take translation invariance (spatial and temporal homogeneity) for granted, and focus more specifically on the Lorentz group.

2.2. Quaternions

The quaternions are numbers that can be written in the form $a + b\,\mathbf{i} + c\,\mathbf{j} + d\,\mathbf{k}$, where a, b, c, and d are real numbers, and $\mathbf{i}$, $\mathbf{j}$, and $\mathbf{k}$ are the quaternion units which satisfy the famous relation

$$\mathbf{i}^2 = \mathbf{j}^2 = \mathbf{k}^2 = \mathbf{ijk} = -1. \tag{2}$$

They form a four–dimensional number system that is generally treated as an extension of the complex numbers. We shall define the quaternion conjugate of the quaternion $\mathbf{q} = a + b\,\mathbf{i} + c\,\mathbf{j} + d\,\mathbf{k}$ to be $\mathbf{q}^\star = a - b\,\mathbf{i} - c\,\mathbf{j} - d\,\mathbf{k}$, and define the norm of $\mathbf{q}$ to be $\mathbf{qq}^\star = |\mathbf{q}|^2 = a^2 + b^2 + c^2 + d^2 \in \mathbb{R}^+$. This allows us to evaluate the quaternion inverse as $\mathbf{q}^{-1} = \mathbf{q}^\star/|\mathbf{q}|^2$.

Trying to define a "norm" as $\mathbf{q}^2 = a^2 - b^2 - c^2 - d^2$, while superficially more "relativistic", violates the usual mathematical definition of "norm", and furthermore is not useful when it comes to evaluating the quaternion inverse $\mathbf{q}^{-1}$.

For current purposes we focus our attention on pure quaternions. That is, we consider quaternions of the form $a\,\mathbf{i} + b\,\mathbf{j} + c\,\mathbf{k}$. Many quaternion operations become much simpler when we are dealing with pure quaternions. For example, the product of two pure quaternions $\mathbf{p}$ and $\mathbf{q}$ is given by $\mathbf{pq} = -\vec{p}\cdot\vec{q} + (\vec{p}\times\vec{q})\cdot(\mathbf{i},\mathbf{j},\mathbf{k})$, where, in general, we shall set $\mathbf{v} = \vec{v}\cdot(\mathbf{i},\mathbf{j},\mathbf{k})$. From this, we obtain the useful relations

$$[\mathbf{p},\mathbf{q}] = 2(\vec{p}\times\vec{q})\cdot(\mathbf{i},\mathbf{j},\mathbf{k}), \quad \text{and} \quad \{\mathbf{p},\mathbf{q}\} = -2\,\vec{p}\cdot\vec{q}. \tag{3}$$

A notable consequence of (3) is $\mathbf{q}^2 = -\vec{q}\cdot\vec{q} = -q^2 = -|\mathbf{q}|^2$. There is a natural isomorphism between the space of pure quaternions and $\mathbb{R}^3$ given by

$$\mathbf{i} \mapsto \hat{x}, \quad \mathbf{j} \mapsto \hat{y}, \quad \mathbf{k} \mapsto \hat{z}; \tag{4}$$

where $\hat{x}$, $\hat{y}$, and $\hat{z}$ are the standard unit vectors in $\mathbb{R}^3$.

One of the most common uses for quaternions today (2020) is in the computer graphics community, where they are used to compactly and efficiently generate rotations in 3-space. Indeed, if $\mathbf{q} = \cos(\theta/2) + \hat{\mathbf{u}}\sin(\theta/2)$ is an arbitrary unit quaternion and $\mathbf{v}$ is the image of a vector in $\mathbb{R}^3$ under the isomorphism (4), then the mapping $\mathbf{v} \mapsto \mathbf{qvq}^{-1}$ rotates $\mathbf{v}$ through an angle θ about the axis defined by $\hat{\mathbf{u}}$. The mapping $\mathbf{v} \mapsto \mathbf{qvq}^{-1}$ is called *quaternion conjugation* by $\mathbf{q}$.

3. Combining Two 3-Velocities

In the paper by Giust, Vigoureux, and Lages [15], see also Reference [16], (and the somewhat related discussion in Reference [10]), a method is developed to compactly combine relativistic velocities in two space dimensions, and by extension, coplanar relativistic velocities in 3 space dimensions. In the following subsection, we first provide a short summary of their approach, and then in the next subsection extend their method to general non-coplanar 3-velocities.

3.1. Velocities in the (x,y)-Plane

The success of this Giust, Vigoureux, and Lages approach relies on the angle addition formula for the hyperbolic tangent function,

$$\tanh(\xi_1 + \xi_2) = \frac{\tanh\xi_1 + \tanh\xi_2}{1 + \tanh\xi_1\tanh\xi_2}. \tag{5}$$

The tanh function is a natural choice for combining relativistic velocities since it is limited to the interval $[-1, 1]$. Indeed, using the rapidity ξ defined by $v = \tanh(\xi)$, we can easily combine collinear relativistic speeds using Equation (5). In order to use this for the combination of non-collinear relativistic 2-velocities, we replace each 2-velocity $\vec{v}$ by the complex number

$$V = \tanh(\xi/2)\,e^{i\varphi}. \tag{6}$$

Here ξ is the rapidity of the velocity $\vec{v}$, and φ gives the orientation of $\vec{v}$ according to some observer in the plane defined by $\vec{v}_1$ and $\vec{v}_2$. Giust, Vigoureux, and Lages then define the composition law $\oplus$ for coplanar velocities $\vec{v}_1$ and $\vec{v}_2$ by

$$W = \tanh\frac{\xi}{2}\,e^{i\varphi_{1\oplus 2}} = V_1 \oplus V_2 = \frac{V_1 + V_2}{1 + \overline{V_2}\,V_1} = \frac{\tanh\frac{\xi_1}{2}\,e^{i\varphi_1} + \tanh\frac{\xi_2}{2}\,e^{i\varphi_2}}{1 + \tanh\frac{\xi_2}{2}\,e^{-i\varphi_2}\,\tanh\frac{\xi_1}{2}\,e^{i\varphi_1}}, \tag{7}$$

where $\overline{V}$ is the standard complex conjugate of V. By using $\xi/2$ instead of ξ in Equations (6) and (7), we are actually dealing with the "relativistic half–velocities", $\tanh(\xi/2)$, (sometimes called the "symmetric velocities"), where

$$w = \tanh(\xi/2); \qquad v = \tanh(\xi) = \frac{2w}{1+w^2}. \tag{8}$$

That is:

$$w = \tanh\left(\frac{1}{2}\,\tanh^{-1}(v)\right) = \frac{v}{1+\sqrt{1-v^2}}. \tag{9}$$

Using Equations (5) and (7) we can easily retrieve the real velocity from the half-velocity by using the $\oplus$ operator: $v = \tanh\xi = \tanh\xi/2 \oplus \tanh\xi/2 = w \oplus w$. In terms of the half velocities

$$w_{1\oplus 2}\,e^{i\varphi_{1\oplus 2}} = \frac{w_1\,e^{i\varphi_1} + w_2\,e^{i\varphi_2}}{1 + w_1 w_2\;e^{i(\varphi_1-\varphi_2)}}. \tag{10}$$

The $\oplus$ addition law is non-commutative, which is most easily seen by first setting $\theta = \varphi_2 - \varphi_1$, then $\Omega = \varphi_{1\oplus 2} - \varphi_{2\oplus 1}$, and finally observing that the ratio

$$e^{i\Omega/2} = \frac{1 + \tanh\frac{\xi_1}{2}\,\tanh\frac{\xi_2}{2}e^{i\theta}}{1 + \tanh\frac{\xi_1}{2}\,\tanh\frac{\xi_2}{2}e^{-i\theta}} = \frac{1 + w_1 w_2 e^{i\theta}}{1 + w_1 w_2 e^{-i\theta}} \tag{11}$$

is not equal to unity for non–zero θ, meaning that $\Omega = \varphi_{1\oplus 2} - \varphi_{2\oplus 1}$ is non-zero.

The angle $\Omega = \varphi_{1\oplus 2} - \varphi_{2\oplus 1}$ is in fact the Wigner angle Ω, so an expression for this angle can be obtained by taking the real and imaginary parts of Equation (11):

$$\tan\frac{\Omega}{2} = \frac{\tanh\frac{\xi_1}{2}\,\tanh\frac{\xi_2}{2}\,\sin\theta}{1 + \tanh\frac{\xi_1}{2}\,\tanh\frac{\xi_2}{2}\,\cos\theta} = \frac{w_1 w_2 \sin\theta}{1 + w_1 w_2 \cos\theta}. \tag{12}$$

This expression does not explicitly appear in Reference [15] though something functionally equivalent, in the form $\Omega = 2\arg(1 + w_1 w_2 e^{i\theta})$, appears in Reference [16].

The $\oplus$ law can be applied to any number of coplanar velocities by iteration:

$$W = (((V_1 \oplus V_2) \oplus \cdots \oplus V_{n-1}) \oplus V_n). \tag{13}$$

Thus it would be desirable to cleanly extend this formalism to general three-dimensional velocities. Note that the order of composition is important, as we shall see in more detail below, the $\oplus$ operation is in general *not* associative.

3.2. General 3-Velocities

We now extend the result of Giust, Vigoureux, and Lages to arbitrary 3-velocities in three dimensions.

3.2.1. Algorithm

Suppose we have a velocity $\vec{v}_i$ in the (x, y)-plane, represented by the pure quaternion $\mathbf{w}_i = \tanh(\xi_i/2)\hat{\mathbf{n}}_i = \tanh(\xi_i/2)\ (\mathbf{i}\cos\theta_i + \mathbf{j}\sin\theta_i)$. Using the rules for quaternion multiplication, we can write this as $\mathbf{w}_i = \tanh(\xi_i/2)\ (\cos\theta_i + \mathbf{k}\sin\theta_i)\mathbf{i}$. The term inside the brackets now looks very similar to what would be a natural extension of the exponential function to the quaternions, $\mathrm{e}^{\mathbf{k}\theta} = \cos\theta + \mathbf{k}\sin\theta$. To formalise this, we define the exponential of a quaternion $\mathbf{q}$ by the power series

$$\mathrm{e}^{\mathbf{q}} = \sum_{k=0}^{\infty} \frac{\mathbf{q}^k}{k!}. \tag{14}$$

To calculate an explicit formula for Equation (14), we first consider the case of a pure quaternion $\mathbf{u}$. We know from Section 2.2 that for a pure quaternion we have $\mathbf{u}^2 = -|\mathbf{u}|^2$, and so we find $\mathbf{u}^3 = -|\mathbf{u}|^2\mathbf{u}$, $\mathbf{u}^4 = |\mathbf{u}|^4$, and so on. Thus, we can compute

$$\mathrm{e}^{\mathbf{u}} \equiv \sum_{k=0}^{\infty} \frac{\mathbf{u}^k}{k!} = \left(1 - \frac{1}{2!}|\mathbf{u}|^2 + \frac{1}{4!}|\mathbf{u}|^4 - \dots\right) + \frac{\mathbf{u}}{|\mathbf{u}|}\left(|\mathbf{u}| - \frac{1}{3!}|\mathbf{u}|^3 + \frac{1}{5!}|\mathbf{u}|^5 - \dots\right) = \cos|\mathbf{u}| + \hat{\mathbf{u}}\sin|\mathbf{u}|. \tag{15}$$

Following the same procedure above, we find the exponential of a pure unit quaternion $\hat{\mathbf{u}}$ and real number ϕ to be

$$\mathrm{e}^{\hat{\mathbf{u}}\phi} = \cos\phi + \hat{\mathbf{u}}\ \sin\phi. \tag{16}$$

This nice result reflects the expression for the exponential of a complex number.

We can now extend this result to any arbitrary quaternion $\mathbf{q} = a + \mathbf{u}$ by noting that the real number a commutes with all the terms in $\mathbf{u}$, thereby allowing us to write $\mathrm{e}^{\mathbf{q}} = \mathrm{e}^a\mathrm{e}^{\mathbf{u}}$, where $\mathrm{e}^{\mathbf{u}}$ has the same form as Equation (15). Explicitly,

$$\mathrm{e}^{\mathbf{q}} = \mathrm{e}^a(\cos|\mathbf{u}| + \hat{\mathbf{u}}\ \sin|\mathbf{u}|). \tag{17}$$

The exponential of a quaternion possesses many of the same properties as the exponential of a complex number. Two particularly useful ones we use below are

$$\left(\mathrm{e}^{\hat{\mathbf{u}}\phi}\right)^\star = \mathrm{e}^{-\hat{\mathbf{u}}\phi} = \cos\phi - \hat{\mathbf{u}}\sin\phi, \quad \text{and} \quad |\mathrm{e}^{\hat{\mathbf{u}}\phi}| = 1. \tag{18}$$

Using these results, we are now justified in writing

$$\mathbf{w}_i = \tanh(\xi_i/2)\,\mathrm{e}^{\mathbf{k}\theta_i}\ \mathbf{i} = w_i\,\mathrm{e}^{\mathbf{k}\theta_i}\ \mathbf{i} \tag{19}$$

for our velocity in the (x, y)-plane.

Building on this result, we now find it appropriate to define the $\oplus$ operator for general 3-velocities, $\mathbf{w}_1 = w_1\hat{\mathbf{n}}_1$ and $\mathbf{w}_2 = w_2\hat{\mathbf{n}}_2$, by the novel formula:

$$\mathbf{w}_{1\oplus 2} = \mathbf{w}_1 \oplus \mathbf{w}_2 = (1 - \mathbf{w}_1\mathbf{w}_2)^{-1}(\mathbf{w}_1 + \mathbf{w}_2). \tag{20}$$

The usefulness of this novel definition is best understood by looking at a few examples.

3.2.2. Example: Parallel Velocities

We consider two parallel velocities $\vec{v}_1$ and $\vec{v}_2$ represented by the quaternions

$$\mathbf{w}_1 = \tanh\frac{\xi_1}{2}\ \hat{\mathbf{n}} \quad \text{and} \quad \mathbf{w}_2 = \tanh\frac{\xi_2}{2}\ \hat{\mathbf{n}}, \tag{21}$$

respectively. Our composition law (20) then gives

$$\begin{aligned}\mathbf{w}_{1\oplus 2} &= \left(1+\tanh\tfrac{\xi_1}{2}\tanh\tfrac{\xi_2}{2}\right)^{-1}\left(\tanh\tfrac{\xi_1}{2}\,\hat{\mathbf{n}}+\tanh\tfrac{\xi_2}{2}\,\hat{\mathbf{n}}\right)\\ &= \frac{\tanh\frac{\xi_1}{2}+\tanh\frac{\xi_2}{2}}{1+\tanh\frac{\xi_1}{2}\tanh\frac{\xi_2}{2}}\,\hat{\mathbf{n}}\\ &= \tanh\left(\tfrac{\xi_1+\xi_2}{2}\right)\hat{\mathbf{n}},\end{aligned} \tag{22}$$

which is equivalent to

$$\mathbf{w}_{1\oplus 2} = \frac{w_1+w_2}{1+w_1 w_2}\,\hat{\mathbf{n}}, \tag{23}$$

and hence, also equivalent to the well–known result for the relativistic composition of two parallel velocities,

$$\vec{v}_1 \oplus \vec{v}_2 = \frac{v_1+v_2}{1+v_1 v_2}\,\hat{n}. \tag{24}$$

3.2.3. Example: Perpendicular Velocities in the x–y Plane

We now consider two perpendicular velocities in the x–y plane. By rotating around the z axis, without loss of generality they can be taken to be given by

$$\mathbf{w}_1 = w_1\mathbf{i}, \quad \mathbf{w}_2 = w_2\,\mathbf{j}, \tag{25}$$

where we have written $\tanh(\xi_1/2) = w_1$ and $\tanh(\xi_2/2) = w_2$ for brevity.

Our composition law then gives a combined velocity of

$$\mathbf{w}_{1\oplus 2} = (1-w_1w_2\mathbf{ij})^{-1}(w_1\mathbf{i}+w_2\mathbf{j}) = \frac{w_1(1-w_2^2)\mathbf{i}+w_2(1+w_1^2)\mathbf{j}}{1+w_1^2w_2^2}, \tag{26}$$

which is definitely not commutative. In contrast the norm is symmetric:

$$|\mathbf{w}_{1\oplus 2}|^2 = \frac{w_1^2(1-w_2^2)^2+w_2^2(1+w_1^2)^2}{(1+w_1^2w_2^2)^2} = \frac{w_1^2+w_2^2+w_1^2w_2^4+w_2^2w_1^4}{(1+w_1^2w_2^2)^2} = \frac{w_1^2+w_2^2}{1+w_1^2w_2^2}. \tag{27}$$

Here the $\mathbf{w}_i$ are the "relativistic half–velocities" $w_i = \tanh(\xi_i/2)$, so the full velocities are

$$|\mathbf{v}_i|^2 = |\mathbf{w}_i \oplus \mathbf{w}_i|^2 = \frac{4w_i^2}{(1+w_i^2)^2}, \tag{28}$$

and so give a final speed of

$$|\mathbf{v}_{1\oplus 2}|^2 = \frac{4\left(w_1^2+w_2^2\right)}{\left(1+w_1^2w_2^2\right)\left[1+\frac{w_1^2+w_2^2}{1+w_1^2w_2^2}\right]^2} = \frac{4(w_1^2+w_2^2)\left(1+w_1^2w_2^2\right)}{\left[\left(1+w_1^2\right)\left(1+w_2^2\right)\right]^2}. \tag{29}$$

The non-quaternionic result for the composition of two perpendicular velocities is [22]

$$||\vec{v}_{1\oplus 2}||^2 = v_1^2+v_2^2-v_1^2v_2^2. \tag{30}$$

Thus, we find

$$||\vec{v}_{1\oplus 2}||^2 = \frac{4w_1^2}{(1+w_1^2)^2}+\frac{4w_2^2}{(1+w_2^2)^2}-\frac{16w_1^2w_2^2}{(1+w_1^2)^2(1+w_2^2)^2} = \frac{4(w_1^2+w_2^2)(1+w_1^2w_2^2)}{\left[(1+w_1^2)(1+w_2^2)\right]^2}. \tag{31}$$

And so our composition law $\oplus$ gives the standard result for the composition of two perpendicular velocities in the x–y plane.

3.2.4. Example: Perpendicular Velocities in General

For general perpendicular velocities $\vec{v}_1$ and $\vec{v}_2$ the easiest way of proceeding is to simply rotate to point $\vec{v}_1$ along the x-axis and $\vec{v}_2$ along the y-axis, and just copy the argument above. If one wishes to be more direct then simply define

$$\mathbf{w}_1 = w_1\, \widehat{\mathbf{w}_1}, \qquad \mathbf{w}_2 = w_2\, \widehat{\mathbf{w}_2}; \qquad \widehat{\mathbf{w}_3} = \widehat{\mathbf{w}_1}\, \widehat{\mathbf{w}_2}. \tag{32}$$

In view of the mutual orthogonality of the vectors $\hat{w}_1$, $\hat{w}_2$, and $\hat{w}_3$, the unit quaternions $(\widehat{\mathbf{w}_1}, \widehat{\mathbf{w}_2}, \widehat{\mathbf{w}_3})$ obey exactly the same commutation relations as $(\mathbf{i}, \mathbf{j}, \mathbf{k})$. Thence

$$\mathbf{w}_{1\oplus 2} = (1 - w_1 w_2 \widehat{\mathbf{w}_1}\widehat{\mathbf{w}_2})^{-1}(w_1\widehat{\mathbf{w}_1} + w_2\widehat{\mathbf{w}_2}) = \frac{w_1(1-w_2^2)\widehat{\mathbf{w}_1} + w_2(1+w_1^2)\widehat{\mathbf{w}_2}}{1 + w_1^2 w_2^2}. \tag{33}$$

This now leads to exactly the same results as above; there was no loss of generality inherent in working in the x–y plane.

3.2.5. Example: Reduction to Giust–Vigoureux–Lages Result in the x–y Plane

It is important to note that our composition law $\oplus$ reduces to the composition law of Giust, Vigoureux, and Lages [15] when dealing with planar velocities in the x–y plane. As above, we define general velocities in the $(\mathbf{i}, \mathbf{j})$-plane by $\mathbf{w}_1 = \tanh(\xi_1/2)\mathrm{e}^{\mathbf{k}\phi_1}\ \mathbf{i}$, and $\mathbf{w}_2 = \tanh(\xi_2/2)\mathrm{e}^{\mathbf{k}\phi_2}\ \mathbf{i}$, then, using our composition law (20), we find

$$\mathbf{w}_{1\oplus 2} = \left(1 - \tanh\frac{\xi_1}{2}\mathrm{e}^{\mathbf{k}\phi_1}\mathbf{i}\ \tanh\frac{\xi_2}{2}\mathrm{e}^{\mathbf{k}\phi_2}\mathbf{i}\right)^{-1}\left(\tanh\frac{\xi_1}{2}\mathrm{e}^{\mathbf{k}\phi_1}\mathbf{i} + \tanh\frac{\xi_2}{2}\mathrm{e}^{\mathbf{k}\phi_2}\mathbf{i}\right). \tag{34}$$

But, noting that $\tanh(\xi_2/2)\mathrm{e}^{\mathbf{k}\phi_2}\mathbf{i} = \tanh(\xi_2/2)\mathbf{i}\,\mathrm{e}^{-\mathbf{k}\phi_2}$ and $\mathbf{i}^2 = -1$, we can re-write this as

$$\mathbf{w}_{1\oplus 2} = \left(1 + \tanh\frac{\xi_1}{2}\mathrm{e}^{\mathbf{k}\phi_1}\tanh\frac{\xi_2}{2}\mathrm{e}^{-\mathbf{k}\phi_2}\right)^{-1}\left(\tanh\frac{\xi_1}{2}\mathrm{e}^{\mathbf{k}\phi_1} + \tanh\frac{\xi_2}{2}\mathrm{e}^{\mathbf{k}\phi_2}\right)\mathbf{i}. \tag{35}$$

Now writing

$$\mathbf{w}_{1\oplus 2} = \tanh(\xi_{1\oplus 2}/2)\,\mathrm{e}^{\mathbf{k}\phi_{1\oplus 2}}\,\mathbf{i} \tag{36}$$

we can cancel out the trailing $\mathbf{i}$, to obtain

$$\tanh\frac{\xi_{1\oplus 2}}{2}\,\mathrm{e}^{\mathbf{k}\phi_{1\oplus 2}} = \left(1 + \tanh\frac{\xi_1}{2}\mathrm{e}^{\mathbf{k}\phi_1}\tanh\frac{\xi_2}{2}\mathrm{e}^{-\mathbf{k}\phi_2}\right)^{-1}\left(\tanh\frac{\xi_1}{2}\mathrm{e}^{\mathbf{k}\phi_1} + \tanh\frac{\xi_2}{2}\mathrm{e}^{\mathbf{k}\phi_2}\right). \tag{37}$$

This expression now only contains $\mathbf{k}$, so everything commutes, and we can write

$$w_{1\oplus 2}\,\mathrm{e}^{\mathbf{k}\phi_{1\oplus 2}} = \frac{w_1\,\mathrm{e}^{\mathbf{k}\phi_1} + w_2\,\mathrm{e}^{\mathbf{k}\phi_2}}{1 + w_1\,\mathrm{e}^{\mathbf{k}\phi_1}\,w_2\,\mathrm{e}^{-\mathbf{k}\phi_2}} \tag{38}$$

which is equivalent to the result of Giust, Vigoureux, and Lages [15].

3.2.6. Example: Composition in General Directions

For general velocities $\vec{v}_1$ and $\vec{v}_2$ the easiest way of proceeding is to simply rotate to put $\vec{v}_1$ and $\vec{v}_2$ in the the x–y plane, and just copy the Giust–Vigoureux–Lages argument [15] above. If one wishes to be more direct then simply define

$$\mathbf{w}_1 = w_1\, \widehat{\mathbf{w}_1}, \qquad \mathbf{w}_2 = w_2\, \widehat{\mathbf{w}_2}; \qquad \widehat{\mathbf{w}_3} = \frac{[\widehat{\mathbf{w}_1},\, \widehat{\mathbf{w}_2}]}{|\,[\widehat{\mathbf{w}_1},\, \widehat{\mathbf{w}_2}]\,|}. \tag{39}$$

As long as $\widehat{\mathbf{w}_1}$ is not parallel to $\widehat{\mathbf{w}_2}$, then $\widehat{\mathbf{w}_3}$ is well defined and perpendicular to both $\widehat{\mathbf{w}_1}$ and $\widehat{\mathbf{w}_2}$. With these definitions one can now write

$$\widehat{\mathbf{w}_2} = \exp(\phi\, \widehat{\mathbf{w}_3})\, \widehat{\mathbf{w}_1}. \tag{40}$$

Then, following the discussion above, we see

$$\mathbf{w}_{1\oplus 2} = (1 + w_1\, w_2\, \mathrm{e}^{-\widehat{\mathbf{w}_3}\phi})^{-1}(w_1\, \widehat{\mathbf{w}_1} + w_2\, \widehat{\mathbf{w}_2}) = (1 + w_1\, w_2\, \mathrm{e}^{-\widehat{\mathbf{w}_3}\phi})^{-1}(w_1 + w_2\, \mathrm{e}^{\widehat{\mathbf{w}_3}\phi})\widehat{\mathbf{w}_1}. \tag{41}$$

From this we can extract

$$w_{1\oplus 2}\, \mathrm{e}^{\widehat{\mathbf{w}_3}\phi_{1\oplus 2}} = (1 + w_1\, w_2\, \mathrm{e}^{-\widehat{\mathbf{w}_3}\phi})^{-1}(w_1 + w_2\, \mathrm{e}^{\widehat{\mathbf{w}_3}\phi}) = \frac{(w_1 + w_2\, \mathrm{e}^{\widehat{\mathbf{w}_3}\phi})}{(1 + w_1\, w_2\, \mathrm{e}^{-\widehat{\mathbf{w}_3}\phi})}. \tag{42}$$

Thence

$$w_{1\oplus 2}\, \mathrm{e}^{\widehat{\mathbf{w}_3}\phi_{1\oplus 2}} = \frac{(w_1 + w_2\, \mathrm{e}^{\widehat{\mathbf{w}_3}\phi})}{(1 + w_1\, w_2\, \mathrm{e}^{-\widehat{\mathbf{w}_3}\phi})}. \tag{43}$$

This finally is a fully explicit result for general velocities $\vec{v}_1$ and $\vec{v}_2$, which is manifestly in agreement with the Giust–Vigoureux–Lages results [15].

3.2.7. Uniqueness of the Composition Law

Finally, we might note that the expression for the composition law (20) is not unique. For example, by considering the power-series of $(1 - \mathbf{w}_1\mathbf{w}_2)^{-1}$, we can re-write Equation (20) as

$$\mathbf{w}_{1\oplus 2} = (1 - \mathbf{w}_1\mathbf{w}_2)^{-1}(\mathbf{w}_1 + \mathbf{w}_2) = \sum_{n=0}^{\infty}(\mathbf{w}_1\mathbf{w}_2)^n(\mathbf{w}_1 + \mathbf{w}_2). \tag{44}$$

But, as $\mathbf{w}_1$ and $\mathbf{w}_2$ are pure quaternions, both $\mathbf{w}_1^2$ and $\mathbf{w}_2^2$ are real numbers, and so commute with $\mathbf{w}_1$ and $\mathbf{w}_2$. Thus,

$$\mathbf{w}_{1\oplus 2} = \sum_{n=0}^{\infty}(\mathbf{w}_1\mathbf{w}_2)^n\mathbf{w}_1 + \sum_{n=0}^{\infty}(\mathbf{w}_1\mathbf{w}_2)^n\mathbf{w}_2 = \mathbf{w}_1\sum_{n=0}^{\infty}(\mathbf{w}_2\mathbf{w}_1)^n + \mathbf{w}_2\sum_{n=0}^{\infty}(\mathbf{w}_2\mathbf{w}_1)^n. \tag{45}$$

Consequently we find that our composition law can also be written as

$$\mathbf{w}_{1\oplus 2} = (\mathbf{w}_1 + \mathbf{w}_2)\sum_{n=0}^{\infty}(\mathbf{w}_2\mathbf{w}_1)^n = (\mathbf{w}_1 + \mathbf{w}_2)(1 - \mathbf{w}_2\mathbf{w}_1)^{-1}. \tag{46}$$

Indeed, one could use either Equation (20) or Equation (46) as the definition of the composition law $\oplus$. Nonetheless, we will stick with the convention given in (20).

3.3. Calculating the Wigner Angle

In this section we obtain an expression for the Wigner angle for general 3-velocities using our composition law (20). Our calculations are obtained using the result that the Wigner angle is the angle between the velocities $\mathbf{w}_{1\oplus 2}$ and $\mathbf{w}_{2\oplus 1}$. We first note

$$|\mathbf{w}_{1\oplus 2}| = |\mathbf{w}_{2\oplus 1}| = |1-\mathbf{w}_1\mathbf{w}_2|^{-1}|\mathbf{w}_1+\mathbf{w}_2| = \frac{|w_1\hat{\mathbf{n}}_1 + w_2\hat{\mathbf{n}}_2|}{|1-w_1w_2\,\hat{\mathbf{n}}_1\hat{\mathbf{n}}_2|}. \tag{47}$$

Thence, setting $\cos\theta = \vec{n}_1\cdot\vec{n}_2$ we explicitly verify

$$|\mathbf{w}_{1\oplus 2}| = |\mathbf{w}_{2\oplus 1}| = \sqrt{\frac{w_1^2+w_2^2+2w_1w_2\cos\theta}{1+w_1^2w_2^2+2w_1w_2\cos\theta}}. \tag{48}$$

Now note that because $|\mathbf{w}_{1\oplus 2}| = |\mathbf{w}_{2\oplus 1}|$ it follows that $(\mathbf{w}_{1\oplus 2})\,(\mathbf{w}_{2\oplus 1})^{-1}$ is a unit norm quaternion. In fact it is related to the Wigner angle by

$$\mathrm{e}^{\mathbf{\Omega}} = (\mathbf{w}_{1\oplus 2})\,(\mathbf{w}_{2\oplus 1})^{-1}. \tag{49}$$

Then

$$\mathrm{e}^{\mathbf{\Omega}} = \Big((1-\mathbf{w}_1\mathbf{w}_2)^{-1}(\mathbf{w}_1+\mathbf{w}_2)\Big)\,\Big((1-\mathbf{w}_2\mathbf{w}_1)^{-1}(\mathbf{w}_2+\mathbf{w}_1)\Big)^{-1} \tag{50}$$

But since for a product of quaternions $(\mathbf{q}_1\mathbf{q}_2)^{-1} = \mathbf{q}_2^{-1}\mathbf{q}_1^{-1}$ this reduces to

$$\mathrm{e}^{\mathbf{\Omega}} = (1-\mathbf{w}_1\mathbf{w}_2)^{-1}(1-\mathbf{w}_2\mathbf{w}_1). \tag{51}$$

Now

$$\mathbf{w}_1\mathbf{w}_2 = -w_1w_2\cos\theta + (\vec{w}_1\times\vec{w}_2)\cdot(\mathbf{i},\mathbf{j},\mathbf{k}). \tag{52}$$

Let us define

$$\hat{\Omega} = \frac{(\vec{w}_1\times\vec{w}_2)}{|\vec{w}_1\times\vec{w}_2|}; \qquad \text{so} \qquad \hat{w}_1\times\hat{w}_2 = \sin\theta\;\hat{\Omega}. \tag{53}$$

Then setting $\hat{\mathbf{\Omega}} = \hat{\Omega}\cdot(\mathbf{i},\mathbf{j},\mathbf{k})$ so that $\mathbf{\Omega} = \Omega\,\hat{\mathbf{\Omega}}$ we have:

$$\mathbf{w}_1\mathbf{w}_2 = -w_1w_2(\cos\theta - \sin\theta\,\hat{\mathbf{\Omega}}) = -w_1w_2\,\mathrm{e}^{-\theta\hat{\mathbf{\Omega}}}. \tag{54}$$

Consequently the Wigner angle satisfies

$$\mathrm{e}^{\mathbf{\Omega}} = \mathrm{e}^{\Omega\hat{\mathbf{\Omega}}} = \left(1+w_1w_2\,\mathrm{e}^{-\theta\hat{\mathbf{\Omega}}}\right)^{-1}\left(1+w_1w_2\,\mathrm{e}^{\theta\hat{\mathbf{\Omega}}}\right) = \frac{1+w_1w_2\,\mathrm{e}^{\theta\hat{\mathbf{\Omega}}}}{1+w_1w_2\,\mathrm{e}^{-\theta\hat{\mathbf{\Omega}}}}. \tag{55}$$

Equivalently,

$$\mathrm{e}^{\Omega\hat{\mathbf{\Omega}}/2} = \frac{1+w_1w_2\,\mathrm{e}^{\theta\hat{\mathbf{\Omega}}}}{|1+w_1w_2\,\mathrm{e}^{\theta\hat{\mathbf{\Omega}}}|}. \tag{56}$$

Taking the scalar and vectorial parts of Equation (56), we finally obtain

$$\tan\frac{\Omega}{2} = \frac{w_1w_2\sin\theta}{1+w_1w_2\cos\theta} = \frac{|\vec{w}_1\times\vec{w}_2|}{1+\vec{w}_1\cdot\vec{w}_2}, \tag{57}$$

as an explicit expression for the Wigner angle Ω.

The simplicity of Equation (57) compared to existing formulae for Ω in the literature, shows how the composition law (20) can lead to much tidier and simpler formulae than other methods allowed for. This can be seen as the extension of the result (12) to more general velocities.

We can write Equation (57) in a perhaps more familiar (though possibly more tedious) form by first noting that from Equation (28) we have

$$w_i = \frac{1-\sqrt{1-v_i^2}}{v_i} = \frac{\gamma_i - 1}{\sqrt{\gamma_i^2-1}} = \sqrt{\frac{\gamma_i-1}{\gamma_i+1}} = \frac{\sqrt{\gamma_i^2-1}}{\gamma_i+1} = \frac{v_i\gamma_i}{\gamma_i+1}, \tag{58}$$

and so

$$\tan\frac{\Omega}{2} = \frac{v_1v_2\gamma_1\gamma_2\sin\theta}{(1+\gamma_1)(1+\gamma_2)+v_1v_2\gamma_1\gamma_2\cos\theta}. \tag{59}$$

We can check two interesting cases of Equation (57) for when $\theta = 0$ (parallel velocities) and when $\theta = \pi/2$ (perpendicular velocities). We can see directly that, for parallel velocities, the associated Wigner angle is given by $\tan(\Omega/2) = 0$, so that $\Omega = n\pi$ for $n \in \mathbb{Z}$; whilst for perpendicular velocities, the associated Wigner angle is simply given by $\tan(\Omega/2) = w_1w_2$.

It is easiest to check our results against the literature using the somewhat messier Equation (59), in which case parallel velocities again give $\tan(\Omega/2) = 0$, whilst perpendicular velocities give

$$\tan(\Omega/2) = \frac{v_1v_2\gamma_1\gamma_2}{(1+\gamma_1)(1+\gamma_2)}, \tag{60}$$

which agrees with the results given in Reference [22].

4. Combining Three 3-Velocities

Let us now see what happens when we relativistically combine 3 half-velocities.
We shall calculate, compare, and contrast $\mathbf{w}_{(1\oplus2)\oplus3}$ with $\mathbf{w}_{1\oplus(2\oplus3)}$.

4.1. Combining 3 Half-Velocities: $\mathbf{w}_{(1\oplus2)\oplus3}$

Start from our key result

$$\mathbf{w}_{1\oplus2} = \mathbf{w}_1 \oplus \mathbf{w}_2 = (1-\mathbf{w}_1\mathbf{w}_2)^{-1}(\mathbf{w}_1+\mathbf{w}_2), \tag{61}$$

and iterate it to yield

$$\mathbf{w}_{(1\oplus2)\oplus3} = \{1-(1-\mathbf{w}_1\mathbf{w}_2)^{-1}(\mathbf{w}_1+\mathbf{w}_2)\mathbf{w}_3\}^{-1}\{(1-\mathbf{w}_1\mathbf{w}_2)^{-1}(\mathbf{w}_1+\mathbf{w}_2)+\mathbf{w}_3\}. \tag{62}$$

It is now a matter of straightforward quaternionic algebra to check that

$$\begin{aligned}\mathbf{w}_{(1\oplus2)\oplus3} &= \{(1-\mathbf{w}_1\mathbf{w}_2)^{-1}(1-\mathbf{w}_1\mathbf{w}_2-(\mathbf{w}_1+\mathbf{w}_2)\mathbf{w}_3)\}^{-1} \\ &\qquad \times\{(1-\mathbf{w}_1\mathbf{w}_2)^{-1}(\mathbf{w}_1+\mathbf{w}_2)+\mathbf{w}_3\} \\ &= (1-\mathbf{w}_1\mathbf{w}_2-(\mathbf{w}_1+\mathbf{w}_2)\mathbf{w}_3)^{-1}(1-\mathbf{w}_1\mathbf{w}_2)\{(1-\mathbf{w}_1\mathbf{w}_2)^{-1}(\mathbf{w}_1+\mathbf{w}_2)+\mathbf{w}_3\} \\ &= (1-\mathbf{w}_1\mathbf{w}_2-(\mathbf{w}_1+\mathbf{w}_2)\mathbf{w}_3)^{-1}\{(\mathbf{w}_1+\mathbf{w}_2)+(1-\mathbf{w}_1\mathbf{w}_2)\mathbf{w}_3\}.\end{aligned} \tag{63}$$

Ultimately we have the novel result

$$\mathbf{w}_{(1\oplus2)\oplus3} = \{1-\mathbf{w}_1\mathbf{w}_2-\mathbf{w}_1\mathbf{w}_3-\mathbf{w}_2\mathbf{w}_3\}^{-1}\{\mathbf{w}_1+\mathbf{w}_2+\mathbf{w}_3-\mathbf{w}_1\mathbf{w}_2\mathbf{w}_3\}. \tag{64}$$

An alternative formulation starts from

$$\mathbf{w}_{1\oplus2} = \mathbf{w}_1 \oplus \mathbf{w}_2 = (\mathbf{w}_1+\mathbf{w}_2)(1-\mathbf{w}_2\mathbf{w}_1)^{-1}, \tag{65}$$

which when iterated yields

$$\mathbf{w}_{(1\oplus2)\oplus3} = \{(\mathbf{w}_1+\mathbf{w}_2)(1-\mathbf{w}_2\mathbf{w}_1)^{-1}+\mathbf{w}_3\}\{1-\mathbf{w}_3(\mathbf{w}_1+\mathbf{w}_2)(1-\mathbf{w}_2\mathbf{w}_1)^{-1}\}^{-1}. \tag{66}$$

Thence a little straightforward quaternionic algebra verifies that

$$\begin{aligned}\mathbf{w}_{(1\oplus 2)\oplus 3} &= \{(\mathbf{w}_1+\mathbf{w}_2)+\mathbf{w}_3(1-\mathbf{w}_2\mathbf{w}_1)\}(1-\mathbf{w}_2\mathbf{w}_1)^{-1}\\ &\qquad\times\{1-\mathbf{w}_3(\mathbf{w}_1+\mathbf{w}_2)(1-\mathbf{w}_2\mathbf{w}_1)^{-1}\}^{-1}\\ &= \{(\mathbf{w}_1+\mathbf{w}_2)+\mathbf{w}_3(1-\mathbf{w}_2\mathbf{w}_1)\}\{(1-\mathbf{w}_2\mathbf{w}_1)-\mathbf{w}_3(\mathbf{w}_1+\mathbf{w}_2)\}^{-1}\end{aligned} \tag{67}$$

Ultimately we have the novel result

$$\mathbf{w}_{(1\oplus 2)\oplus 3}=\{\mathbf{w}_1+\mathbf{w}_2+\mathbf{w}_3-\mathbf{w}_3\mathbf{w}_2\mathbf{w}_1\}\{1-\mathbf{w}_2\mathbf{w}_1-\mathbf{w}_3\mathbf{w}_1-\mathbf{w}_3\mathbf{w}_2\}^{-1}. \tag{68}$$

So we have found two equivalent and novel formulae for $\mathbf{w}_{(1\oplus 2)\oplus 3}$, Equations (64) and (68).

4.2. Combining 3 Half-Velocities: $\mathbf{w}_{1\oplus(2\oplus 3)}$

In contrast, the situation for $\mathbf{w}_{1\oplus(2\oplus 3)}$ is considerably more subtle. Start from the key result that

$$\mathbf{w}_{2\oplus 3}=\mathbf{w}_2\oplus\mathbf{w}_3=(1-\mathbf{w}_2\mathbf{w}_3)^{-1}(\mathbf{w}_2+\mathbf{w}_3), \tag{69}$$

and iterate it to yield

$$\mathbf{w}_{1\oplus(2\oplus 3)}=\{1-\mathbf{w}_1(1-\mathbf{w}_2\mathbf{w}_3)^{-1}(\mathbf{w}_2+\mathbf{w}_3)\}^{-1}\{\mathbf{w}_1+(1-\mathbf{w}_2\mathbf{w}_3)^{-1}(\mathbf{w}_2+\mathbf{w}_3)\}. \tag{70}$$

The relevant quaternionic algebra is now a little trickier

$$\begin{aligned}\mathbf{w}_{1\oplus(2\oplus 3)} &= \{1-\mathbf{w}_1(1-\mathbf{w}_2\mathbf{w}_3)^{-1}(\mathbf{w}_2+\mathbf{w}_3)\}^{-1}(1-\mathbf{w}_2\mathbf{w}_3)^{-1}\\ &\qquad\times\{(1-\mathbf{w}_2\mathbf{w}_3)\mathbf{w}_1+(\mathbf{w}_2+\mathbf{w}_3)\}\\ &= \{(1-\mathbf{w}_2\mathbf{w}_3)(1-\mathbf{w}_1(1-\mathbf{w}_2\mathbf{w}_3)^{-1}(\mathbf{w}_2+\mathbf{w}_3)\}^{-1}\\ &\qquad\times\{(1-\mathbf{w}_2\mathbf{w}_3)\mathbf{w}_1+(\mathbf{w}_2+\mathbf{w}_3)\}\\ &= \{1-\mathbf{w}_2\mathbf{w}_3-(1-\mathbf{w}_2\mathbf{w}_3)\mathbf{w}_1(1-\mathbf{w}_2\mathbf{w}_3)^{-1}(\mathbf{w}_2+\mathbf{w}_3)\}^{-1}\\ &\qquad\times\{\mathbf{w}_1+\mathbf{w}_2+\mathbf{w}_3-\mathbf{w}_2\mathbf{w}_3\mathbf{w}_1\}.\end{aligned} \tag{71}$$

To proceed we note that

$$\begin{aligned}(1-\mathbf{w}_2\mathbf{w}_3)\mathbf{w}_1(1-\mathbf{w}_2\mathbf{w}_3)^{-1} &= \left(\frac{1-\mathbf{w}_2\mathbf{w}_3}{|1-\mathbf{w}_2\mathbf{w}_3|}\right)\mathbf{w}_1\left(\frac{1-\mathbf{w}_2\mathbf{w}_3}{|1-\mathbf{w}_2\mathbf{w}_3|}\right)^{-1}\\ &= e^{-\mathbf{\Omega}_{2\oplus 3}/2}\,\mathbf{w}_1\,e^{+\mathbf{\Omega}_{2\oplus 3}/2}.\end{aligned} \tag{72}$$

Thence we have the novel result

$$\mathbf{w}_{1\oplus(2\oplus 3)}=\{1-\mathbf{w}_2\mathbf{w}_3-(e^{-\mathbf{\Omega}_{2\oplus 3}/2}\mathbf{w}_1e^{+\mathbf{\Omega}_{2\oplus 3}/2})(\mathbf{w}_2+\mathbf{w}_3)\}^{-1}\{\mathbf{w}_1+\mathbf{w}_2+\mathbf{w}_3-\mathbf{w}_2\mathbf{w}_3\mathbf{w}_1\}. \tag{73}$$

While structurally similar to the formulae (64) and (68) for $\mathbf{w}_{(1\oplus 2)\oplus 3}$ the present result (73) for $\mathbf{w}_{1\oplus(2\oplus 3)}$ is certainly different—the Wigner angle $\mathbf{\Omega}_{2\oplus 3}$ now makes an explicit appearance, also the form of the triple-product $\mathbf{w}_2\mathbf{w}_3\mathbf{w}_1$ is different.

4.3. Combining 3 Half-Velocities: (Non)-Associativity

From (64) and (68) for $\mathbf{w}_{(1\oplus 2)\oplus 3}$, and (73) for $\mathbf{w}_{1\oplus(2\oplus 3)}$, it is clear that relativistic composition of velocities is in general not associative. (See for instance the discussion in References [28,29], commenting on Reference [29].) A sufficient condition for associativity, $\mathbf{w}_{(1\oplus 2)\oplus 3}=\mathbf{w}_{1\oplus(2\oplus 3)}$, is to enforce

$$e^{-\mathbf{\Omega}_{2\oplus 3}/2}\mathbf{w}_1e^{+\mathbf{\Omega}_{2\oplus 3}/2}=\mathbf{w}_1;\qquad\text{and}\qquad \mathbf{w}_1\mathbf{w}_2\mathbf{w}_3=\mathbf{w}_2\mathbf{w}_3\mathbf{w}_1. \tag{74}$$

That is, a sufficient condition for associativity is

$$[\boldsymbol{\Omega}_{2\oplus 3}, \mathbf{w}_1] = 0; \qquad \text{and} \qquad [\mathbf{w}_1, \mathbf{w}_2\mathbf{w}_3] = 0. \tag{75}$$

But note $\boldsymbol{\Omega}_{2\oplus 3} \propto [\mathbf{w}_2, \mathbf{w}_3]$ and $\mathbf{w}_2\mathbf{w}_3 = \frac{1}{2}\{\mathbf{w}_2, \mathbf{w}_3\} + \frac{1}{2}[\mathbf{w}_2, \mathbf{w}_3]$. Since $\{\mathbf{w}_2, \mathbf{w}_3\} \in \mathbb{R}$, we then have $[\mathbf{w}_1, \mathbf{w}_2\mathbf{w}_3] = \frac{1}{2}[\mathbf{w}_1, [\mathbf{w}_2, \mathbf{w}_3]]$. This now implies that these two sufficiency conditions are in fact identical; so a sufficient condition for associativity is

$$[\mathbf{w}_1, [\mathbf{w}_2, \mathbf{w}_3]] = 0. \tag{76}$$

This sufficient condition for associativity can also be written as the vanishing of the vector triple product

$$\vec{w}_1 \times (\vec{w}_2 \times \vec{w}_3) = 0. \tag{77}$$

Equivalently

$$\vec{v}_1 \times (\vec{v}_2 \times \vec{v}_3) = 0. \tag{78}$$

4.4. Specific Non-Coplanar Example

As a final example of the power of the quaternion formalism, let us consider a specific intrinsically non-coplanar example. Let $\mathbf{w}_1 = w_1\mathbf{i}$, $\mathbf{w}_2 = w_2\,\mathbf{j}$, and $\mathbf{w}_3 = w_3\mathbf{k}$ be three mutually perpendicular half-velocities. (So this configuration does automatically satisfy the associativity condition discussed above.) Then we have already seen that:

$$\mathbf{w}_1 \oplus \mathbf{w}_2 = \frac{w_1(1-w_2^2)\mathbf{i} + w_2(1+w_1^2)\mathbf{j}}{1+w_1^2w_2^2}; \qquad w_{1\oplus 2}^2 = \frac{w_1^2+w_2^2}{1+w_1^2w_2^2}. \tag{79}$$

Furthermore, since $\mathbf{w}_1 \oplus \mathbf{w}_2$ is perpendicular to $\mathbf{w}_3$, we have

$$(\mathbf{w}_1 \oplus \mathbf{w}_2) \oplus \mathbf{w}_3 = \frac{w_{1\oplus 2}(1-w_3^2)\hat{\mathbf{n}}_{1\oplus 2} + w_3(1+w_{1\oplus 2}^2)\mathbf{k}}{1+w_{1\oplus 2}^2w_3^2}, \tag{80}$$

and

$$w_{(1\oplus 2)\oplus 3}^2 = \frac{w_{(1\oplus 2)}^2 + w_3^2}{1+w_{(1\oplus 2)}^2w_3^2} = \frac{w_1^2+w_2^2+w_3^2+w_1^2w_2^2w_3^2}{1+w_1^2w_2^2+w_2^2w_3^2+w_3^2w_1^2}. \tag{81}$$

A little algebra now yields the manifestly non-commutative result

$$(\mathbf{w}_1 \oplus \mathbf{w}_2) \oplus \mathbf{w}_3 = \frac{(1-w_2^2)(1-w_3^2)\mathbf{w}_1 + (1+w_1^2)(1-w_3^2)\mathbf{w}_2 + (1+w_1^2)(1+w_2^2)\mathbf{w}_3}{1+w_1^2w_2^2+w_2^2w_3^2+w_3^2w_1^2}. \tag{82}$$

In this particular case we can also explicitly show that

$$(\mathbf{w}_1 \oplus \mathbf{w}_2) \oplus \mathbf{w}_3 = \mathbf{w}_1 \oplus (\mathbf{w}_2 \oplus \mathbf{w}_3), \tag{83}$$

though (as discussed above) associativity fails in general.

5. Conclusions

Herein we have provided a simple, elegant, and novel algebraic method for combining special relativistic 3-velocities using quaternions:

$$\mathbf{w}_{1\oplus 2} = \mathbf{w}_1 \oplus \mathbf{w}_2 = (1-\mathbf{w}_1\mathbf{w}_2)^{-1}(\mathbf{w}_1+\mathbf{w}_2) = (\mathbf{w}_1+\mathbf{w}_2)(1-\mathbf{w}_2\mathbf{w}_1)^{-1}. \tag{84}$$

The construction also leads to a simple, elegant, and novel formula for the Wigner angle:

$$e^{\Omega} = e^{\Omega\,\hat{\Omega}} = (1 - \mathbf{w}_1\mathbf{w}_2)^{-1}(1 - \mathbf{w}_2\mathbf{w}_1), \tag{85}$$

in terms of which

$$\hat{\mathbf{w}}_{1\oplus 2} = e^{\Omega/2}\, \frac{\mathbf{w}_1 + \mathbf{w}_2}{|\mathbf{w}_1 + \mathbf{w}_2|}; \qquad \hat{\mathbf{w}}_{2\oplus 1} = e^{-\Omega/2}\, \frac{\mathbf{w}_1 + \mathbf{w}_2}{|\mathbf{w}_1 + \mathbf{w}_2|}. \tag{86}$$

All of the non-commutativity associated with non-collinearity of 3-velocities is automatically and rather efficiently dealt with by the quaternion algebra.

Author Contributions: Conceptualization, T.B. and M.V.; methodology, T.B. and M.V.; software, T.B. and M.V.; validation, T.B. and M.V.; formal analysis, T.B. and M.V.; resources, M.V.; writing—original draft preparation, T.B. and M.V.; writing—review and editing, T.B. and M.V.; supervision, M.V.; project administration, M.V.; funding acquisition, M.V. All authors have read and agreed to the published version of the manuscript.

Funding: T.B. was supported by a Victoria University of Wellington MSc scholarship, and was also indirectly supported by the Marsden Fund, via a grant administered by the Royal Society of New Zealand. M.V. was directly supported by the Marsden Fund, via a grant administered by the Royal Society of New Zealand.

Conflicts of Interest: The authors declare no conflict of interest.

References

1. Silberstein, L. Quaternionic form of relativity. *Philisophical Mag.* **1912**, *23*, 790–809. [CrossRef]
2. Silberstein, L. *The Theory of Relativity*; Macmillan and Co: London, UK, 1914
3. Silberstein, L. Available online: https://en.wikipedia.org/wiki/Ludwik_Silberstein (accessed on 8 December 2020).
4. Dirac, P.A.M. Application of quaternions to Lorentz transformations. *Proc. R. Ir. Acad.* **1944**, *50*, 261–270.
5. Rastall, P. Quaternions in Relativity. *Rev. Mod. Phys.* **1964**, *36*, 820. [CrossRef]
6. Girard, P.R. The quaternion group and modern physics. *Eur. J. Phys.* **1984**, *5*, 25–32. [CrossRef]
7. Ungar, A.A. The relativistic velocity composition paradox and the Thomas rotation. *Found. Phys.* **1989**, *19*, 1385–1396. [CrossRef]
8. Mocanu, C.I. On the relativistic velocity composition paradox and the Thomas rotation. *Found. Phys. Lett.* **1992**, *5*, 443–456. [CrossRef]
9. De Leo, S. Quaternions and Special Relativity. *J. Math. Phys.* **1996**, *37*, 2955–2968. [CrossRef]
10. Friedman, Y. *Physical Applications of Homogeneous Balls*; Progress in Mathematical Physics; Springer: Birkhäuser, Basel, 2005; Volume 40, doi:10.1007/978-0-8176-8208-8. [CrossRef]
11. Friedman, Y.; Semon, M.D. Relativistic acceleration of charged particles in uniform and mutually perpendicular electric and magnetic fields as viewed in the laboratory frame. *Phys. Rev. E* **2005**, *72*, 026603. [CrossRef]
12. Greiter, M.; Schuricht, D. Imaginary in all directions: An Elegant formulation of special relativity and classical electrodynamics. *Eur. J. Phys.* **2003**, *24*, 397–401. [CrossRef]
13. Yefremov, A.P. Theory of relativity in quaternion spinors. *Gravit. Cosmol.* **2016**, *22*, 97–106. [CrossRef]
14. Wigner, E. On unitary representations of the inhomogeneous Lorentz group. *Ann. Math.* **1939**, *40*, 149–204. [CrossRef]
15. Giust, R.; Vigoureux, J.-M.; Lages, J. Generalized composition law from 2 × 2 matrices. *Am. J. Phys.* **2009**, *77*, 1068–1073. [CrossRef]
16. Lages, J.; Giust, R.; Vigoureux, J.-M. Composition law for polarizers. *Phys. Rev.* **2008**, *78*, 033810. [CrossRef]
17. Thomas, L.H. The motion of the spinning electron. *Nature* **1926**, *117*, 514–514. [CrossRef]
18. Fisher, G.P. Thomas precession. *Am. J. Phys.* **1972**, *40*, 1772. [CrossRef]
19. Ferraro, M.; Thibeault, R. Generic composition of boosts: An elementary derivation of the Wigner rotation. *Eur. J. Phys.* **1999**, *20*, 143. [CrossRef]
20. Malykin, G.B. Thomas precession: Correct and incorrect solutions. *Phys. Uspekhi* **2006**, *49*, 837–853. [CrossRef]
21. Ritus, V.I. On the difference between Wigner's and Møller's approaches to the description of Thomas precession. *Phys. Uspekhi* **2007**, *50*, 95–101. [CrossRef]

22. O'Donnell, K.; Visser, M. Elementary analysis of the special relativistic combination of velocities, Wigner rotation, and Thomas precession. *Eur. J. Phys.* **2011**, *32*, 1033–1047. [CrossRef]
23. Achilles, R.; Bonfiglioli, A. The early proofs of the theorem of Campbell, Baker, Hausdorff, and Dynkin. *Arch. Hist. Exact Sci.* **2012**, *66*, 295–358. [CrossRef]
24. Goldberg, K. The formal power series for $\log(e^x e^y)$. *Duke Math. J.* **1956**, *23*, 13–21. [CrossRef]
25. Van-Brunt, A.; Visser, M. Special-case closed form of the Baker-Campbell-Hausdorff formula. *J. Phys. A* **2015**, *48*, 225207. [CrossRef]
26. Van-Brunt, A.; Visser, M. Simplifying the Reinsch algorithm for the Baker-Campbell-Hausdorff series. *J. Math. Phys.* **2016**, *57*, 023507. [CrossRef]
27. Van-Brunt, A.; Visser, M. Explicit Baker-Campbell-Hausdorff Expansions. *Mathematics* **2018**, *6*, 135. [CrossRef]
28. Ungar, A.A. Thomas precession: A kinematic effect of the algebra of Einstein's velocity addition law. Comments on 'Deriving relativistic momentum and energy: II. Three-dimensional case'. *Eur. J. Phys.* **2006**, *27*, L17. [CrossRef]
29. Sonego, S.; Pin, M. Deriving relativistic momentum and energy: II. Three-dimensional case (CORRIGENDUM). *Eur. J. Phys.* **2005**, *26*, 851–856, Correction in **2006**, *27*, 685. [CrossRef]

universe

MDPI

Article

Making a Quantum Universe: Symmetry and Gravity

Houri Ziaeepour [1,2]

[1] Institut UTINAM, CNRS UMR 6213, Observatoire de Besançon, Université de Franche Compté, 41 bis ave. de l'Observatoire, BP 1615, 25010 Besançon, France; hz@mssl.ucl.ac.uk or houriziaeepour@gmail.com

[2] Mullard Space Science Laboratory, University College London, Holmbury St. Mary, Dorking GU5 6NT, UK

Received: 05 September 2020; Accepted: 17 October 2020; Published: 23 October 2020

Abstract: So far, none of attempts to quantize gravity has led to a satisfactory model that not only describe gravity in the realm of a quantum world, but also its relation to elementary particles and other fundamental forces. Here, we outline the preliminary results for a model of quantum universe, in which gravity is fundamentally and by construction quantic. The model is based on three well motivated assumptions with compelling observational and theoretical evidence: quantum mechanics is valid at all scales; quantum systems are described by their symmetries; universe has infinite independent degrees of freedom. The last assumption means that the Hilbert space of the Universe has $SU(N \to \infty) \cong$ area preserving Diff.(S_2) symmetry, which is parameterized by two angular variables. We show that, in the absence of a background spacetime, this Universe is trivial and static. Nonetheless, quantum fluctuations break the symmetry and divide the Universe to subsystems. When a subsystem is singled out as reference—observer—and another as clock, two more continuous parameters arise, which can be interpreted as distance and time. We identify the classical spacetime with parameter space of the Hilbert space of the Universe. Therefore, its quantization is meaningless. In this view, the Einstein equation presents the projection of quantum dynamics in the Hilbert space into its parameter space. Finite dimensional symmetries of elementary particles emerge as a consequence of symmetry breaking when the Universe is divided to subsystems/particles, without having any implication for the infinite dimensional symmetry and its associated interaction-percived as gravity. This explains why gravity is a universal force.

Keywords: quantum gravity; quantum mechanics; symmetry; quantum cosmology

1. Introduction and Summary of Results

More than a century after the discovery of general relativity and description of gravitational force as the modification of spacetime geometry by matter and energy, we still lack a convincing model for explaining these processes in the framework of quantum mechanics. Appendix A briefly reviews the history of efforts for finding a consistent Quantum Gravity (QGR) model. Despite tremendous effort of generations of scientists, none of proposed models presently seem fully satisfactory.

Quantization of gravity is inevitable. Examples of inconsistencies in a universe where matter is ruled by quantum mechanics but gravity is classical are well known [1,2]. In addition, in [2], it is argued that there must be an inherent relation between gravity and quantum mechanics. Otherwise, the universality of Planck constant $\hbar$ as quantization scale cannot be explained.[1] Aside from these arguments, the fact that there is no fundamental mass/energy scale in quantum mechanics means that it has to have a close relation with gravity that provides a dimensionful fundamental constant, namely the Newton gravitational constant G_N or equivalently the Planck mass $M_P \equiv \sqrt{\hbar c/G_N}$, where c is the

[1] Nonetheless, Ref. [3] advocates a context dependent Planck constant.

speed of light in vacuum (or equivalently Planck length scale $L_P \equiv \hbar/cM_P$). We should remind that a dimensionful scale does not arise in conformal or scale independent models. Indeed, conformal symmetry is broken by gravity, which provides the only fundamental dimensionful constant to play the role of a ruler and make distance and mass measurements meaningful.[2]

In what concerns the subject of this volume, namely representations of inhomogeneous Lorentz symmetry (called also Poincaré group), they were under special interest since decades ago, hoping that they help formulate gravity as a renormalizable quantum field. The similarity of the compact group of local Lorentz transformations to Yang–Mills gauge symmetry has encouraged quantum gravity models that are based on the first order formulation of general relativity. These models use vierbein formalism and extension of gauge group of elementary particles to accommodate Poincaré group [5,6]. However, Coleman–Madula theorem [7] on S-matrix symmetries—local transformations of interacting fields that asymptotically approach Poincaré symmetry at infinity—invalidates any model in which Poincaré and internal symmetries are not factorized. According to this theorem total symmetry of a grand unification model, including gravity, must be a tensor product of spacetime and internal symmetries. Otherwise, the model must be supersymmetric [8] or VEV of the gauge field should not be flat [9]. However, we know that even if supersymmetry is present at M_P scale, it is broken at low energies. Moreover, any violation of Coleman–Madula theorem and Lorentz symmetry at high energies can be convoyed to low energies [10] and violate e.g., equivalence principle and other tested predictions of general relativity [11,12]. For these reasons, modern approaches to the unification of gravity as a gauge field with other interactions consider the two sectors as separate gauge field models. In addition, in these models gravity sector usually has topological action to make the formulation independent of the geometry of underlying spacetime, see e.g., [13–15]. However, like other quantum gravity candidates these models suffer from various issues. The separation of internal and gravitational gauge sectors means that these models are not properly speaking a grand unification. Moreover, similar to other approaches to QGR, these models do not clarify the nature of spacetime, its dimensionality, and relation between gravity and internal symmetries.

In addition to consistency with general relativity, cosmology, and particle physics, a quantum model unifying gravity with other forces is expected to solve well known problems that are related to gravity and spacetime, such as: physical origin of the arrow of time; apparent information loss in black holes; and, UV and IR singularities in Quantum Field Theory (QFT) and general gravity.[3] There are also other issues that a priori should be addressed by a QGR model, but are less discussed in the literature:

1. **Should spacetime be considered as a physical entity similar to quantum fields associated to particles, or rather it presents a configuration space ?**

   General relativity changed spacetime from a rigid entity to a deformable media. However, it does not specify whether spacetime is a physical reality or a property of matter, which ultimately determines its geometry and topology. We remind that in the framework of QFT vacuum is not the empty space of classical physics, see e.g., [16,17]. In particular, in the presence of gravity the naive definition of quantum vacuum is frame dependent. A frame-independent definition exists [18] and it is very far from classical concept of an

[2] In addition to M_P, we need two other fundamental constants to describe physics and cosmology: the Planck constant $\hbar$ and maximum speed of information transfer that experiments show to be the speed of light in classical vacuum. We remind that triplet constants $(\hbar, c, M_P)$ are arbitrary and can take any nonzero positive value. The selection of their values amounts to the definition of a system of units for measuring other physical quantities. In QFT literature usually $\hbar = 1$ and $c = 1$ are used. In this system of units—called high energy physics units [4]—$\hbar$ and c are dimensionless.

[3] Some quantum gravity models such as loop quantum gravity emphasize the quantization of gravity alone. However, giving the fact that gravity is a universal force and interacts with matter and other forces, its quantization necessarily has impact on them. Therefore, any quantum gravity only model would be, at best, incomplete.

empty space. Explicitly or implicitly, some of models reviewed in Appendix A address this question.

2. **Is there any relation between matter and spacetime ?**

   In general relativity matter modifies the geometry of spacetime, but the two entities are considered as separate and stand alone. In string theory spacetime and matter fields—compactified internal space—are considered and treated together, and spacetime has a physical reality that is similar to matter. By contrast, many other QGR candidates only concentrate their effort on the quantization of spacetime and gravitational interaction. Matter is usually added as an external ingredient and it does not intertwine in the construction of quantum gravity and spacetime.

3. **Why do we perceive the Universe as a three-dimensional (3D) space (plus time) ?**

   None of extensively studied quantum gravity models discussed in Appendix A answer this question, despite the fact that it is the origin of many troubles for them. For instance, the enormous number of possible models in string theory is due to the inevitable compactification of extra-dimensions to reduce the dimension of space to the observed 3 + 1. In background independent models, the dimension of space is a fundamental assumption and essential for many technical aspects of their construction. In particular, the definition of Ashtekar variables [19] for $SU(2) \cong SO(3)$ symmetry and its relation with spin foam description of loop quantum gravity [20] are based on the assumption of a 3D real space. On the other hand, according to holography principle, the maximum amount of information that is containable in a quantum system is proportional to its area rather than volume. If the information is projected and available on the boundary, it is puzzling why we should perceive the volume.

In a previous work [21], we advocated the foundational role of symmetries in quantum mechanics and reformulated its axioms accordingly, see Appendix B for a summary. Of course, the crucial role of symmetries in quantum systems is well known. However, axioms of quantum mechanics à la Dirac and von Neumann consider an abstract Hilbert space and do not specify its relation with symmetries of quantum systems. In addition to symmetries of their classical Lagrangian, Hilbert space of quantum systems represents $SU(N)$ group, called state symmetry, see Appendix C for more details. Transformation of states by this group modifies their coherence, and recently quantification of this property and its usefulness as a resource has become a subject of interest in quantum information theory literature [22,23].

Inspired by these developments, in this work we study a standalone quantum system, which is considered to be the Universe.

1.1. Summary of the Model and Results

The model assumes infinite number of independent and simultaneously commuting observables in the Universe, but no background spacetime. Hilbert space $\mathcal{H}_U$ of such system represents $SU(N \to \infty)$ symmetry. However, in absence of a background spacetime, its dynamics is trivial and its Lagrangian is defined on the group manifold of $SU(\infty)$ symmetry. Therefore, states are pure gauge. The vector space of gauge transformations, corresponding to linear transformations of the Hilbert space, is $\mathcal{B}[\mathcal{H}_U] \cong SU(\infty)$. On the other hand, quantum fluctuations break the state symmetry and factorize the Hilbert space to blocks of tensor product of subspaces according to criteria studied in [24,25]. For each subsystem, the rest of the Universe plays the role of a background parameterized by three continuous quantities that can be identified with the classical space. Moreover, division of the Universe to subsystems leads to emergence of time and its arrow à la Page & Wootters [26] or similar methods [27]. We show that the 3 + 1 dimensional parameter space is, in general, curved and invariant under inhomogeneous Lorentz transformations and its curvature is determined by quantum states of the subsystems. We also comment on the signature of parameter space metric. Based on these observations, we interpret $SU(\infty)$ sector of the model as Quantum Gravity. The finite rank factorized symmetries become local gauge fields acting on a Hilbert space that presents matter fields.

These results demonstrate the importance of the division of Universe to subsystems and the distinction of observer and clock from the rest. Nonetheless, in contrast to the Copenhagen interpretation of quantum mechanics, the absence of observer does not make the model meaningless, but trivial and static. This model answers some of issues raised in questions 1–3 raised earlier. In particular, it clarifies the nature of spacetime and its dimensionality, and provides an explanation for the universality of gravitational force.

A crucial proposal of the model is that what we perceive as classical spacetime is the configuration (parameter) space of its content. In other words, rather than saying particles/objects (such as strings) live in a 3 + 1 dimensional space, according to this model we can say that an ensemble of abstract objects with $SU(\infty) \times G$ symmetry look like a 3 + 1 dimensional infinite curved spacetime with gravity, where subsystems are fields that represent group G as a local gauge symmetry. Thus, we can completely neglect the geometric interpretation and just consider the Universe as an infinite tensor product. This aspect of the model is similar to the approach of [28]. However, their model is somehow inverse of that studied here. They use tensor product and quantum entanglement to make a symplectic geometry that becomes a continuous curved spacetime when the number of tensor product factors approaches to infinity. The drawback is that symplectic geometries defined by graphs can be embedded in any space of dimension $D \geqslant 2$. Consequently, they cannot explain the dimension of the spacetime.

Axioms and structure of the model is discussed in Section 2. Lagrangian of the system before its division is described in Section 3. Properties of the model after symmetry breaking and division of the Universe are studied in Section 4. Section 5 presents a brief comparison of this model with string and loop quantum gravity. Section 6 presents outlines and prospective for future investigations. Accompanying appendices contain technical details and review of previous results. Appendix A gives a short recount of the history of quantum gravity models. Appendix B summaries the axioms of quantum mechanics in symmetry language. State space and its associated symmetry is reviewed in Appendix C. Properties of $SU(\infty)$ and its representations are summarized in Appendix D and its Cartan decomposition in Appendix E.

2. An Infinite Quantum Universe

Our departure point for constructing a quantum universe consists of three well motivated assumptions with compelling observational and theoretical evidence:

I. Quantum mechanics is valid at all scales and applies to every entity, including the Universe as a whole;
II. Any quantum system is described by its symmetries and its Hilbert space represents them;
III. The Universe has an infinite number of independent degrees of freedom.

The last assumption means that the Hilbert space of the Universe $\mathcal{H}_U$ is infinite dimensional and represents the group $SU(\infty)$. There is sufficient evidence in favour of such an assumption. For instance, the thermal distribution of photons at IR limit contains an infinite number of quanta with energies approaching zero and there is no minimum energy limit. For this reason, vacuum can be considered to be a superposition of multi-particle states of any type—not just photons—without any limit on their number [18]. In general relativity, there is no upper limit for gravitons wavelength and thereby their number. Of course, one may argue that a lower limit on energy or spacetime volume may exist. Nonetheless, for any practical application the number of subsystems/quanta in the Universe can be considered to approach infinity. Indeed, even in quantum gravity models that assume a symplectic structure for spacetime, such as spin foam/loop quantum gravity and causal sets, there is not a fixed lattice of spacetime and the number of spacetime states is effectively infinite.

The algebra that is associated to the $SU(\infty)$ coherence (state) symmetry of the above model is defined as[4]:

$$[\hat{L}_a, \hat{L}_b] = \frac{\hbar}{cM_P} f^c_{ab} \hat{L}_c = L_P f^c_{ab} \hat{L}_c \tag{1}$$

where operators $\hat{L}_\alpha \in \mathcal{B}[\mathcal{H}_U]$ are generators of algebra $su(\infty)$ and f^c_{ab} are its structure coefficients. They are normalized such that the r.h.s. of (1) explicitly depends on the Planck constant $\hbar$. If $\hbar \to 0$, the r.h.s. becomes null, and the algebra becomes abelian and homomorphic to $\bigotimes^{N\to\infty} U(1)$ — in agreement with the symmetry of configuration space of classical systems, explained in Appendix C. The same happens if $M_P \to \infty$, that is when Planck mass scale is much larger than scale of interest. In both cases $L_P \to 0$. Assuming that $SU(\infty)$ symmetry of the Universe can be associated to gravitational interaction—we will provide more evidence in favour of this claim later—the above limits mean that, in both cases, gravity becomes negligible.[5]

It is well known that $\mathcal{B}[\mathcal{H}_U] \cong SU(\infty) \cong$ area preserving Diff(S_2) [29,30], where S_2 is 2D sphere. In fact, $SU(\infty)$ is homomorphic to area preserving diffeomorphism of any two-dimensional (2D) Riemann surface [31–33]. Therefore, here S_2 can be any 2D surface, rather than just sphere. This theorem can be heuristically understood as the following: any compact 2D Riemann surface can be obtained from sphere by removing a measure zero set of pairs of points and sticking the rest of the surface pair-by-pair together. Although surfaces with different genus are topologically different, they are homomorphic. This property may be important in the presence of subsystems with singularity, such as black holes, in which part of the parameter space is inaccessible. From now on, we call a 2D surface that its diffeomorphism represents $SU(\infty)$ a diffeo-surface.

Homomorphism between $SU(\infty)$ and Diff(S_2) makes it possible to expand $\hat{L}_a$'s with respect to spherical harmonic functions, depending on angular coordinates (θ, ϕ) on a sphere. Moreover, owing to the Cartan decomposition, $SU(\infty)$ generators can be described as a tensor product of Pauli matrices [29,34]. In this case, indices in (1) consist of a pair $(l, m) \mid l = 0, \cdots, \infty; -l \leqslant m \leqslant +l$. Appendix E reviews decomposition and indexing of $SU(\infty)$ generators. We continue to use single letters for the indices of generators when there is no need for their explicit description.

The algebra (1) is not enough to make the system quantic and as usual $\hat{L}_a$'s must respect Heisenberg commutation relations:

$$[\hat{L}_a, \hat{J}_b] = -i\delta_{ab}\hbar. \tag{2}$$

where $\hat{J}_a \in \mathcal{B}[\mathcal{H}^*_U]$ is the dual of $\hat{L}_a$ and $\mathcal{H}^*_U$ is the dual Hilbert space of the Universe. As there is a one-to-one correspondence between $\hat{L}$'s and $\hat{J}$'s, they satisfy the same algebra, represent the same symmetry group, namely $SU(\infty)$, and they have their own expansion to spherical harmonics. Owing to $SU(\infty) \cong$ Diff(S_2), vectors of the Hilbert space are differentiable complex functions of angular coordinates (θ, ϕ). Thus, spherical harmonic functions constitute an orthogonal basis for $\mathcal{H}_U$. The Cartan subalgebra of $\mathcal{B}[\mathcal{H}_U] \cong SU(\infty)$ is also infinite dimensional.

The quantum Universe defined here is static, because there is no background space or time in the model. Nonetheless, in Section 4, we show that continuous degrees of freedom similar to space and time naturally arise when the Universe is divided to subsystems. The short argument goes as the following:

We assume that eigen states of the Hilbert space of the Universe are not abstract objects and physically exist. This assumption is supported by the fact that in Standard Model (SM) of particle physics states that constitute a basis for its Hilbert space and for its space of linear transformations are indeed observed particles (fields). Consequently, taking into account the assumption that $\mathcal{H}_U$ is infinite

4 In this work, all vector spaces and algebras are defined on complex number field $\mathbb{C}$, unless explicitly mentioned otherwise.

5 Although in (1) we show the dimensional scale $\hbar/M_P$ in the definition of operators and their algebra, for the sake of convenience in the rest of this work, we include it in the operators, except when its explicit presentation is necessary for the discussion.

dimensional, we conclude that the Universe must consist of infinite number of particles/subsystems. Although subsystems may have some common properties, which make them indistinguishable from each others, there are many other distinguishable aspects, which discriminate them from each others. This statement is in agreement with the corollary presented in Appendix B regarding the divisibility of a quantum Universe, derived from axioms of quantum mechanics. Thus, this conclusion is independent of details of the model.

Notice that, without the assumption about physical existence of eigen states, an infinite dimensional Hilbert space does not necessarily mean Universe must be infinitely divisible. Hilbert space of many quantum systems have an infinite number of states. However, they do not necessarily occur in each instance (copy) of the system. The case of a Universe is different, because, by definition, there is only one copy of it. Therefore, every eigen state of a complete basis of its Hilbert space must physically exist. Otherwise, it can be completely discarded.

In the next sections, we make this argument more rigorous and explain how it can lead to a 3 + 1 dimensional spacetime and internal gauge symmetry of elementary particles. We begin with constructing a Lagrangian for this static model and show that it is trivial.

3. Lagrangian of the Universe

Although the infinite dimensional Universe described in the previous section is static, it has to satisfy constraints imposed by symmetries associated to it. They are analogous to constraints imposed on systems in thermodynamic equilibrium. Although there is no time variation in such systems, a priori small perturbations occur, for instance, by absorption and emission of energy. They must be in balance with each others, otherwise the system would lose its equilibrium. Therefore, it is useful to define a Lagrangian that quantifies these constraints. In the case of present model, the Lagrangian should quantify $SU(\infty)$ symmetry and its representation by $\mathcal{H}_U$.

Lagrangian of a system must be invariant under transformations of fields by application of members of its symmetry group. As there is no background spacetime in this model, the most appealing candidate is a Lagrangian similar to Yang–Mills, but without a kinetic term. In such a situation, the only available quantities are invariants of the symmetry group:

$$\mathcal{L}_U = \int d^2\Omega \sqrt{|g^{(2)}|} \left[\frac{1}{2} \sum_{a,b} L_a^*(\theta,\phi) L_b(\theta,\phi) \text{tr}(\hat{L}_a \hat{L}_b) + \frac{1}{2} \sum_a L_a \text{tr}(\hat{L}_a \rho(\theta,\phi)) \right], \qquad d^2\Omega \equiv d(\cos\theta) d\phi \quad (3)$$

where g is the determinant of 2D metric of the diffeo-surface. If we use description (A5) for $\hat{L}$ operators, $a = b = 1$. If we use (A12) expression, $a, b = (l, m)$, $l = 0, \cdots, \infty; -m \leqslant l \leqslant +m$. The latter case explicitly demonstrates the Cartan decomposition of $SU(\infty)$ to $SU(2)$ factors, as described in Appendix E. Notice that (θ, ϕ) are internal variables [30], reflecting the fact that vectors of the Hilbert space representing $SU(\infty)$ are functions on a 2D Riemann surface. For the same reason, in contrast to usual Lagrangians in QFT, there is no term containing derivatives with respect to these parameters in $\mathcal{L}_U$. If we use differential representation of $\hat{L}_{lm}$ defined in (A5) and apply it to amplitudes $L_{lm}(\theta, \phi)$, the first term in the Lagrangian will depend on the partial derivatives of amplitudes, just like in the QFT. However, it is straightforward to see that derivatives with respect to $\cos\theta$ and ϕ will have different amplitudes and, thereby, the kinetic term will be unconventional and non-covariant, unless we consider amplitudes $L_{lm}(\theta, \phi)$ as functions of the metric of a deformed sphere. This is the explicit demonstration of $SU(\infty) \cong \text{Diff}(S_2)$ invariance of this Lagrangian.

Generators $\hat{T}_a, \hat{T}_b \in SU(N), \forall N$ can be normalized, such that $\text{tr}(T_a T_b) \propto \delta_{ab}$, see e.g., [29]. In analogy with field strength in Yang–Mills theories, the function $L_a(\theta, \phi)$ can be interpreted as the amplitude of the contribution of operator $\hat{L}_a$ in the dynamics of the Universe. Due to global $U(1)$ symmetry of operators applied to a quantum state, L_a's are, in general, complex. On the other hand, when considering the Cartan decomposition of $SU(\infty)$ to tensor product of $SU(2)$ factors and the fact that $\sigma^\dagger = (\sigma*)^t = \sigma$, we conclude that $\hat{L}_a^\dagger = \hat{L}_a$, Similar to QFT, one can use $\mathcal{L}_U$ to define a path integral. In the absence of time, the path integral presents the excursion of states in the Hilbert space by successive application of $\hat{L}_a$ operators. Nonetheless, owing to $SU(\infty)$ symmetry, variation of states is equivalent to gauge transformation and non-measurable.

The analogy of $\mathcal{L}_U$ with Yang–Mills theory has interesting consequences. For instance, differential representation of $\hat{L}_{lm}$ defined in (A5) can be written as $\hat{L}_{lm} = \sqrt{|g^{(2)}|}\epsilon^{\mu\nu}(\partial_\mu Y_{lm})\partial_\nu$. In classical limit, one can consider that $\hat{L}_{lm}$ acts on the field amplitude L_{lm} and the first term in the integrand of Lagrangian $\mathcal{L}_U$ can be arranged, such that it becomes proportional to Ricci scalar $R^{(2)}$. As the geometry of 2D diffeo-surface is arbitrary, for each set of L_{lm} the metric $g_{\mu\nu}$ can be chosen, such that L_{lm} dependent part of the integrand becomes proportional to Ricci scalar for that metric. Thus, in classical limit the first term is topological.[6] We could arrive to this conclusion inversely. Because $SU(\infty) \cong \text{Diff}(S_2)$, in the classical limit the Lagrangian should be the same as Einstein gravity in a static 2D curve space. Thus, the first term in (3) can be replaced by $\int d^2\Omega\sqrt{|g^{(2)}|}R^{(2)}$. Then, the definition of $\hat{L}_{lm}$ operators in (A5) and amplitudes L_{lm} can be used to write $R^{(2)}$ with respect to $\hat{L}_{lm}$ and relate metric and connection of the 2D surface to amplitudes Ł$_{lm}$. We leave a detailed demonstration of these relations to a future work. The relation between gauge field term in $\mathcal{L}_U$ and Riemann curvature in classical limit is crucial for interpretation of this term as gravity when the Universe is divided to subsystems.

Notice that, in both representations of $SU(\infty)$, namely Cartan decomposition to tensor product of $SU(2)$ factors and diffeomorphism of 2D surfaces, angular coordinates θ and ϕ play the role of parameters that identify/index the members of the symmetry group. Consequently, their quantization is meaningless. This is consistent with interpretation of Einstein equation as an equation of state [35]. Presuming the physical reality of Hilbert space and operators applied to it, as discussed in the previous section and in Appendix C, we can interpret L_{lm} as intensity of force mediator particles related to the symmetry represented by operators $\hat{L}_{lm}$, and ρ in the second term of the Lagrangian $\mathcal{L}_U$ as density matrix of matter.

Although $\mathcal{L}_U$ is static, we can apply a variational principle with respect to amplitudes to obtain field equations and find equilibrium values of L_{lm} and ρ. However, it is easily seen that solutions of these equations are trivial. At equilibrium $L_{lm} \to 0$ and $\rho_{lm} \to 0$, see Appendix E.2 for the details. Because $SU(\infty)^n \cong SU(\infty)\ \forall\ n$, this solution has properties of a frame independent vacuum of a many-particle Universe defined by using coherent states [18]. Their similarity implicitly implies that the Universe is divisible and consists of infinite number of particles/subsystems interacting through mediator particles of $SU(\infty)$ force, which is the action of $\hat{L}_{lm}$. We investigate this conclusion in more details in the next section.

4. Division to Subsystems

There are many ways to see that the quantum vacuum (equilibrium) solution of a Universe with $\mathcal{L}_U$ Lagrangian (3) is not stable. Of course there are quantum fluctuations. They are nothing else than random application of $\hat{L}_{lm}$ operators, in other words random scattering of force mediator quanta by matter. They project the Hilbert space to itself. However, owing to $SU(\infty)$ symmetry of Lagrangian, states are globally equivalent and the Universe maintain its equilibrium. Nonetheless, locally states are different and they do not respond to $\hat{L}_{lm}$ in the same manner. Here, locality means restriction of Lagrangian and projections to a subspace of the Hilbert space [25]. As state space is homomorphic to the space of smooth functions on the sphere $f(\theta, \phi)$, the restriction of transformations to a subspace is equivalent to a local deformation of the diffeo-surface. Moreover, the difference between structure coefficients of $SU(\infty)$ can be used to define a locality or closeness among operators that belong to

[6] We remind that $\int_{\mathcal{M}} d^2\Omega\sqrt{|g^{(2)}|}R^{(2)} = 4\pi\chi(\mathcal{M})$, where χ is the Euler characteristic of the compact Riemann 2D surface $\mathcal{M}$. Moreover, Ricci scalar alone does not determine Riemann curvature tensor $R_{\mu\nu}$ and only provides one constraint for three independent components of the metric tensor.

$\mathcal{B}[\mathcal{H}_U]$. These observations are additional evidence to the argument given at the end of Section 3 in favour of the divisibility of the quantum Universe introduced in Section 2 to multi-particle/subsystems.

A quantum system that is divisible to separate and distinguishable subsystems [7] must fulfill 3 conditions [24]:

- There must exist sets of operators $\{A_i\} \subset \mathcal{B}[\mathcal{H}]$ such that $\forall \bar{a} \in \{A_i\}$ and $\forall \bar{b} \in \{A_j\}$, and $i \neq j$, $[\bar{a}, \bar{b}] = 0$;
- Operators in each set $\{A_i\}$ must be local[8];
- $\{A_i\}$'s must be complementarity, which is $\otimes_i \{A_i\} \cong \text{End}(\mathcal{B}[\mathcal{H}])$.

The most trivial way of fulfilling these conditions is a reducible representation of symmetries by $\mathcal{B}[\mathcal{H}]$. In the case of $\mathcal{B}[\mathcal{H}_U] \cong SU(\infty)$, as:

$$SU(\infty)^n \cong SU(\infty) \; \forall \, n \tag{4}$$

the above condition can be easily realized. Moreover, instabilities, quantum correlations, and entanglement may create local symmetries among groups of states and/or operators. There are many examples of such grouping and induced symmetries in many-body systems, see e.g., [36] for a review. A hallmark of induced symmetry by quantum correlations is the formation of anyon quasi-particles having non-abelian symmetry in the fractional quantum Hall effect [37]. On the other hand, there is only one state in the infinite dimensional Hilbert space, in which all pointer states have the same probability, namely the maximally coherent state defined in (A2). Even if a many-body system begins in such a maximally symmetric state, quantum fluctuations rapidly change it to a less coherent and more asymmetric one. In addition, due to (4), irreducible representations of $SU(\infty)$ are partially entangled [25] and there is high probability of clustering of subspaces in a randomly selected state.

Lets assume that such groupings indeed have occurred in the early Universe and they continue to occur at Planck scale. They provide the necessary conditions for division of the Universe to parts or particles with $SU(\infty) \times G \cong SU(\infty)$ as their symmetry. The local symmetry G is assumed to be a compact Lie group of finite rank and respected by all subsystems. Although different subsystems may have different internal symmetries, without a lack of generality, we can assume that G is their tensor product, but some species of particles/subsystems are in singlet representation of some of the component groups.

As the rank of G is assumed to be finite, complementarity condition dictates that the number of subsystems must be infinite to account for the infinite rank of $\mathcal{H}_U$. If states are in a finite dimensional representations of G, at least one of the representations must have infinite multiplicity and their Hilbert space would be infinite dimensional. Thus, despite the division of $\mathcal{B}[\mathcal{H}_U]$, $SU(\infty)$ remains a symmetry of subsystems and $\{A_i\} \subset \mathcal{B}[\mathcal{H}_i] \cong SU(\infty)$, where $\mathcal{H}_i$ is the Hilbert space of subsystem i. Clustering of states and subsystems are usually the hallmark of strong interaction and quantum correlation [36]. Therefore, the interaction of subsystems through internal G symmetry is expected to be stronger than through $SU(\infty)$, thereby the weak gravitation conjecture [38] is satisfied.

We could also formulate the above Universe in a bottom-up manner. Consider the ensemble of infinite number of quantum systems—particles—each having finite symmetry G and coherently mixed with each others. Their ensemble generates a Universe with $SU(\infty) \times G \cong SU(\infty)$ as symmetry represented by its Hilbert space. Therefore, top-down or bottom-up approaches to an infinitely

---

7 In statistical quantum or classical mechanics distinguishability of particles usually means being able to say, for instance, whether it was particle 1 or particle 2 which was observed. Here by distinguishability we mean whether a particle/subsystem can be experimentally detected, i.e., through application of $\hat{L}_{lm}$ to a subspace of parameter space and identified in isolation from other subsystems or the rest of the Universe.

8 This condition is defined for quantum systems in a background spacetime. In the present model there is not such a background. Nonetheless, as explained earlier, locality on the diffeo-surface can be projected to $\mathcal{B}[\mathcal{H}_U]$.

divisible Universe give the same result. The bottom-up view helps to better understand the origin of $SU(\infty)$ symmetry. It shows that, for each subsystem, it is the presence of other infinite number of subsystems and its own interaction with them that is seen as a $SU(\infty)$ symmetry.

4.1. Properties of an Infinitely Divided Quantum Universe

The division of the Universe to subsystems has several consequences. First of all, the global $U(1)$ symmetry of $\mathcal{H}_U$ becomes local, because Hilbert spaces of subsystems $\mathcal{H}_i$, $\forall i$, where index i runs over all subsystems, acquire their own phase symmetry. Therefore, we expect that there is at least one unbroken $U(1)$ local—gauge—symmetry in nature. It may be identified as $U(1)$ symmetry of the Standard Model. From now on, we include this $U(1)$ to the internal symmetry of subsystems G. Additionally, the infinite number of subsystems in the Universe means that each of them has its own representation of $SU(\infty)$ symmetry. However, these representations are not isolated and are part of the $SU(\infty)$ symmetry of the whole Universe. This property is similar to finite intervals on a line, which are homomorphic to $R^{(1)}$ and, at the same time, part of it and have the same algebra. Therefore, the memory of being part of the whole Universe is not washed out with the division to subsystems. Otherwise, according to the corollary discussed in Appendix B subsystems could be considered as separate and isolated universes.

The area of diffeo-surface is irrelevant when only one $SU(\infty)$ is considered. However, it becomes relevant and observable when it is compared with its counterparts for other subsystems. More precisely, homomorphism between Hilbert spaces of two subsystems s and s' defined as:

$$\mathcal{R}_{ss'} : \mathcal{H}_s \rightarrow \mathcal{H}_{s'} \tag{5}$$

can be considered as an additional parameter that is necessary for their identification and indexing. A more qualitative description of how a third continuous parameter emerges from division of Universe to subsystem is given in the next subsection.

4.2. Parameterization of Subsystems

There are various ways to see that the division of the Universe to subsystems defined in Section 2 induces a new continuous parameter. As discussed in the previous subsection, each subsystem represents $SU(\infty) \times G$. When $SU(\infty)$ representation of different subsystems are compared, e.g., through a morphism, the radius of diffeo-surface becomes relevant, because different radius means different area. This dependence allows for classifying subsystems according to a size scale. More precisely, in the definition of $\hat{L}_{lm}$ in (A5), $Y_{lm} \propto r^l$, where r is the distance to centre in spherical coordinates when the 2D surface is embedded in $R^{(3)}$. If we factorize r-dependence part of Y_{lm}, the algebra of $\hat{L}_{lm}$ defined in (A4) becomes:

$$[\hat{L}_{lm}, \hat{L}_{l'm'}]\bigg|_{r=1} = r^{l''-l'-l} f^{l''m''}_{lm,l'm'} \hat{L}_{l''m''}\bigg|_{r=1} \tag{6}$$

where all $\hat{L}_{lm}$ operators are defined for $r = 1$ (in an arbitrary unit). Equation (6) shows that r can be interpreted as a coupling that quantifies the strength of correlation between $\hat{L}_{lm}$ operators. Moreover, due to homomorphism (4), $\hat{L}_{lm}$'s of subsystems are part of $\hat{L}_{lm}$'s of the full system. Consequently, subsystems are never completely isolated and they interact through an algebra similar to (6), but their r factors can be different:

$$[\hat{L}^{(r)}_{lm}, \hat{L}^{(r')}_{l'm'}]\bigg|_{r=1} = r''^{l''} r'^{-l'} r^{-l} f^{l''m''}_{lm,l'm'} \hat{L}^{r''}_{l''m''}\bigg|_{r=1} \tag{7}$$

where r indices on $\hat{L}_{lm}$ operators are added to indicate that they may belong to different subsystems. Nonetheless, the algebra remains the same, because operators $\hat{L}_{lm}$ belong, at the same time, to the global

$SU(\infty)$. On the other hand, the nonlocality of this algebra in the point of view of subsystems should induce a dependence on derivative with respect to parameters when infinitesimal transformations are considered, e.g., in the Lagrangian. Specifically, we expect a relation between r'' and (r, r'), determined by homomorphism (5). In the infinitesimal limit, the r.h.s. of (7) becomes Lie derivative of $\hat{L}_{lm}$ in the direction of $\hat{L}_{l'm'}$ in the manifold that is defined by parameters (r, θ, ϕ, t), where the last parameter is time with respect to an observer, as described in the next subsection.

In summary, after the division of the Universe to subsystems, their $SU(\infty)$ symmetries are indexed by angular parameters (θ, ϕ) and an additional continuous parameter $r = (0, \infty)$. They share the algebra of global $SU(\infty)$, but acquire a new index and, in this sense, their algebra becomes nonlocal. Notably, in the infinitesimal limit the algebra can be considered as the Lie derivative of $\hat{L}_a \in \mathcal{B}[\mathcal{H}]$ operators on the manifold of parameter space $(r, theta, \phi, t)$. Differential properties of the model need more investigation and will be reported elsewhere.

Finally, we can define a conjugate set of parameters for the dual Hilbert space $\mathcal{H}^*_U$ and dual operators $\hat{J}_a$ defined in (2). Therefore, in contrast to some quantum gravity candidates, this model does not have a preference for position or momentum spaces.

4.3. Clocks and Dynamics

The last step for construction of a dynamical quantum Universe is the introduction of a clock by using comparison between variation of states of two subsystems, tagged as system and clock, under the application of operators $\hat{L}_\alpha \in SU(\infty) \times G$ by a third subsystem, tagged as observer, who plays the role of a reference. The necessity of an observer/reference is consistent with the foundation of quantum mechanics, as described in [21]. In the context of the present model, this discrimination can be understood as the following: although the global $SU(\infty)$ symmetry means that any variation of full state by application of $\hat{L}_\alpha$ is a gauge transformation, a variation of subsystems with respect to each others is meaningful and can be quantified.

The technical details of introducing a clock and relative time in quantum mechanics are intensively studied in the literature, see e.g., [27] for a review and proof of the equivalence of different approaches. Here, we describe this procedure through an example. Consider the application of operators $\hat{L}_c \in \mathcal{B}[\mathcal{H}_C]$ and $\hat{L}_s \in \mathcal{B}[\mathcal{H}_s]$ to two subsystems, called clock and system, respectively, such that:

$$\hat{L}_c \rho_c \hat{L}_c^\dagger = \rho_c + d\rho_c = d\rho'_c, \qquad \hat{L}_s \rho_s \hat{L}_s^\dagger = \rho_s + d\rho_s = d\rho'_s \tag{8}$$

Because these operations are local and restricted to subsystems, they are not gauged out. One way of associating a c-number quantity to these variations is to define parameter t, such that, for instance, $dt \equiv |\text{tr}\,(\rho'_c \hat{O}_c) - \text{tr}\,(\rho_c \hat{O}_c)|$, where $\hat{O}_c$ is an observable of the clock subsystem. This quantity is positive and, by definition, incremental. The Hamiltonian operator of the system $H_s \in \mathcal{B}[\mathcal{H}_s]$ according to this clock would be an operator for which $d\rho_s/dt = -i/\hbar[H_s, \rho_s]$.

More generally, defining a clock is equivalent to comparing excursion path of two subsystems in their respective Hilbert space under successive application of $\hat{L}_c$ and $\hat{L}_s$ to them, respectively. The arrow of time arises because through the common $SU(\infty)$ symmetry any operation—even a local one—is communicated to the whole Universe. Thus, inverting the arrow of time amounts to performing an inverse operation on all subsystems, which is extremely difficult. Therefore, although the dynamical equation of one system may be locally symmetric with respect to time reversal, due to global effect of every operation, its effect cannot be easily reversed.

4.4. Geometry of Parameter Space

The final stride of time definition brings the dimension of continuous parameter space necessary for describing states and dynamics of an infinite dimensional divisible Universe to 3 + 1, namely

(r, θ, ϕ, t).[9] Although these parameters arise from different properties of the Universe, namely (θ, ϕ) from $SU(\infty)$ symmetry, r from division to infinite number of subsystems, and t from their relative variation, they are mixed through the global $SU(\infty)$ symmetry, arbitrariness of the choice of reference frame and clock, and quantum superposition of states. Therefore, geometry of the parameter space is $R^{(3+1)}$.

The 2D parameter space of the whole Universe is, by definition, diffeomorphism invariant, as it is the representation of $SU(\infty)$. However, at this stage it is not clear whether the subdivided Universe is rigid, that is only invariant only under global frame transformations of the (3 + 1)D parameter space, or deformable and invariant under its diffeomorphism. Here we show that it is indeed diffeomorphism invariant. Moreover, its geometry is determined by states of subsystems.[10]

Consider a set of 2D diffeo-surfaces representing $SU(\infty)$ symmetries of subsystem. These diffeomorphism can be obtained from application of $\hat{L}_{lm} \in SU(\infty)$ operators to vacuum state of each subsystem, considered to be a sphere. They are smooth functions of parameters (r, θ, ϕ) and can be identified with states of the subsystems, which are also smooth functions of the same parameters. After ordering these surfaces—for instance, according to their average distances [11]—and defining a projection between neighbours [12], such that if on ith surface the point (θ_i, ϕ_i, r_i) is projected to $(\theta_{i+1}, \phi_{i+1}, r_{i+1})$ on $(i+1)$th surface, the distance in $R^{(3)}$ between points in an infinitesimal surface $\Delta\Omega_i < \epsilon^2$ containg (θ_i, ϕ_i, r_i) and infinitesimal surface $\Delta\Omega_{i+1} < \epsilon'^2$ containg $(\theta_{i+1}, \phi_{i+1}, r_{i+1})$ approaches zero if $\epsilon, \epsilon' \to 0$. The path connecting closest points on $\Delta\Omega_i$ and $\Delta\Omega_{i+1}$ defines an orthogonal direction in a deformed $S_2 \times R^{(1)} \cong R^{(3)}$ and the Riemann curvature of this space can be determined from sectional curvature. Therefore, parameter space (or equivalently Hilbert space) is curved. Moreover, as the projection between $\Delta\Omega_i$ and $\Delta\Omega_{i+1}$ used for this demonstration is arbitrary, we conclude that the parameter space is not rigid and its diffeomorphism does not change the physics. The same procedure can be applied when a clock is chosen. Therefore, the above conclusions apply to the full (3 + 1)D parameter space of the subdivided Universe.

Finally, from homomorphism between diffeo-surfaces and states of subsystems, we conclude that (3 + 1)D classical spacetime can be interpreted as parameter space of the Hilbert space of subsystems of the Universe, and gravity as the interaction that is associated to $SU(\infty)$ symmetry.

4.5. Metric Signature

Up to now we indicated the dimension of spacetime—parameter space of $SU(\infty)$ symmetry—after subdivision of the Universe as 3 + 1. This implicitly means that we have considered a Lorentzian metric with negative signature. In special and general relativity, the signature of metric is dictated by observation of the constant speed of light in classical vacuum. Indeed, diffeomorphism invariance, Einstein equation, and interpretation of gravity as curvature of spacetime are independent of signature of the spacetime metric.

---

9 Evidently, in addition to 3 + 1 external parameters each subsystem represents the internal symmetry G, where its representations have their own parameters.

10 Notice that even in classical general relativity diffeomorphism and relation between geometry and state of matter are independent concepts. In particular, Einstein equation is not the only possible relation and a priori other diffeomorphism invariant relations between geometry and matter are allowed—but constrained by experiments.

11 More generally, any measure of difference between states, such as Fubini–Study metric or fidelity can be used to order states. As Hilbert spaces of quantum systems with $SU(\infty)$ symmetry consist of continuous functions, we can use usual analytical tools for defining a distance. However, we should not forget that functions are vectors of a Hilbert space. Moreover, Hilbert space vectors are, in general, complex functions and each projection between diffeo-surfaces corresponds to two projection in the Hilbert space, one for real part and one for imaginary part of vectors.

12 This projection is isomorphic to a homomorphism between $\mathcal{B}[\mathcal{H}_s]$ of subsystems.

In quantum mechanics, Heisenberg uncertainty relation imposes Mandelstam–Tamm constraint [39] on the minimum time necessary for the transition of a quantum state ρ_1 to another perfectly distinguishable state ρ_2 [23]:

$$\Delta t \geqslant \frac{\hbar}{\sqrt{2}} \frac{\cos^{-1} A(\rho_1, \rho_2)}{\sqrt{Q(\rho_1, \hat{H})}},$$
$$A(\rho_1, \rho_2) \equiv \mathrm{tr}(\sqrt{\rho_1}\sqrt{\rho_2}), \qquad Q(\rho, \hat{H}) \equiv \frac{1}{2}|\mathrm{tr}([\sqrt{\rho}, \hat{H}]^2)| \tag{9}$$

where $\hat{H}$ is the system's Hamiltonian. Consider ρ_1 as the state of Universe after selecting and separating an observer and a clock and ρ_2 as an infinitesimal variation of ρ_1, that is $\rho_2 = \rho_1 + d\rho_1$. We assume that the clock is chosen, such that, in (9), minimum time is achieved. Subsequently, (9) becomes:

$$Q(\hat{H}, \rho_1)dt^2 = \mathrm{tr}(\sqrt{d\rho_1}\sqrt{d\rho_1}^\dagger \equiv ds^2 \tag{10}$$

This equation is similar to a Riemann metric for a system at rest with respect to the chosen coordinates frame for the parameter space (r, θ, ϕ, t). On the other hand, the r.h.s. of (10) only depends on the variation of state and is independent of the chosen frame for parameters. Therefore, ds is similar to an infinitesimal separation. A coordinate transformation, i.e., $(r, \theta, \phi, t) \to (r', \theta', \phi', t')$ does not change state of the Universe and is equivalent to a basis transformation in the Hilbert space. On the other hand, the l.h.s. of (10) changes. Considering the similarity of (10) to metric equation, we can write ds^2 as:

$$g_{00}dt'^2 \pm g_{ii}dx'^i dx'^i = ds^2 \tag{11}$$

where we have used Cartesian coordinates in place of spherical. We have chosen parameter transformation such that $g_{0i} = g_{i0} = 0$. We have also assumed $g_{ij} > 0$ and factorized the sign of spatial part of the metric. In these new coordinates, the Hamiltonian associated to the new clock t' is $\hat{H}'$ and Mandelstam–Tamm relation imposes:

$$Q'(\hat{H}', \rho_1)dt'^2 \equiv g_{00}dt'^2 \geqslant \mathrm{tr}(\sqrt{d\rho_1}\sqrt{d\rho_1}^\dagger) = ds^2 \tag{12}$$

For $ds^2 \geqslant 0$, constraint (12) is only satisfied if the sign of spatial part in (11) and thereby the signature of the metric is negative. We remind that Mandelstam–Tamm constraint does not apply to states that do not fulfill distinguishability condition. In these cases, $ds^2 < 0$ is allowed. In classical view of spacetime, they correspond to spacelike events, where two events/states are not causally related. In quantum mechanics, this can be related to nonlocality [40] and the absence of strict causality. Additionally, the explicit dependence of separation on the density matrix in (10) and (12) and its independence of the coordinate frame of the parameters/spacetime confirms and completes the discussion of Section 4.4 regarding the curved geometry of parameter space and diffeomorphism invariance of subdivided Universe.

4.6. Lagrangian of Subsystems

Finally, the Lagrangian of the Universe after the division to subsystems and selection of reference observer and clock takes the following form:

$$\begin{aligned}\mathcal{L}_{U_s} = & \int d^4x\sqrt{-g}\Bigg[\frac{1}{16\pi G_N\hbar}\sum_{l,m,l',m'}\text{tr}(L^*_{lm}(x)L_{l'm'}(x)\hat{L}_{lm}\hat{L}_{l'm'}) + \\ & \frac{1}{8(\pi G_N\hbar)^{1/2}}\Bigg(\sum_{l,m,a}\text{tr}(L_{lm}(x)T_a(x)\hat{L}_{lm}\otimes\hat{T}_a) + \sum_{lm}L_{lm}\text{tr}(\hat{L}_{lm}\otimes\mathbb{1}_G\rho(x))\Bigg) + \\ & \frac{1}{4}\sum_{a,b}\text{tr}(T^*_a(x)T_b(x)\hat{T}_a\hat{T}_b) + \frac{1}{2}\sum_a T_a\text{tr}(\mathbb{1}_{SU(\infty)}\otimes\hat{T}_a\rho(x))\Bigg].\end{aligned} \tag{13}$$

The terms of this Lagrangian can be interpreted as the following. The first term is the Lagrangian for an ensemble of $SU(\infty)$ symmetries of all subsystems, except observer and clock. Amplitudes $L_{l'm'}(x)$ depend on full $SU(\infty)$ parameter space, which is (r, θ, ϕ, t). Due to the nonlocal algebra (7), we expect that $L_{l'm'}(x)$'s include derivative terms. Additionally, L_{lm}'s are normalized such that the usual gravitational coupling be explicit. We notice that, if $\hbar G_N \propto \hbar^2/M_P^2 \rightarrow 0$, the first and the third terms will be canceled. Therefore, the naive classical limit of the model does not include these gravity related terms.

The second term presents gravitational interaction of internal gauge fields and matter, respectively. The third and forth terms together correspond to the Lagrangian of pure gauge fields for local G symmetry and its interaction with matter field. They take the standard form of Yang–Mills models if $T_a(x)$ fields are two-forms in the (3 + 1)D parameter space.

We leave explicit description of $L_{l'm'}(x)$'s and T_a's as functionals of spacetime, and determination of semi-classical limit of the Lagrangian for future works. Nonetheless, the Lagrangian (13) is not completely abstract. L_{lm} operators can be expressed as a tensor product of Pauli matrices and regrouped by r and t indices, which have no other role than associating a group of matrices to subsystems. This is because the tensor product of $SU(\infty)$ is homomorphic to itself. However, such expansion is not very useful and practical for analytical calculations, in particular for finding semi-classical limit of the model.

5. Comparison with Other Quantum Gravity Models

It is useful to compare this model with string theory and Loop Quantum Gravity (LQG)—the two most popular quantum gravity candidates.

A common aspect of string/superstring theories with the present model is the presence of a 2D manifold in their foundation. However, in contrast to string theories, in which a 2D world sheet is introduced as an axiom without any observational support, the presence of a 2D manifold here is a consequence of the infinite symmetry of the Universe, which has compelling observational support. Moreover, the 2D nature of the underlying Universe manifests itself only when the Universe is considered as a whole. Otherwise, it is always perceived as a (3 + 1)D continuum (plus parameters of internal symmetry of subsystems).

In string theory, matter and spacetime are fields living on the 2D world sheet, or equivalently the world sheet can be viewed as being embedded in a multi-dimensional, partially compactified space without any explanation for the origin of such non-trivial structures. On the other hand, in the present model the approach to matter is rather bottom-up. The Cartan decomposition of $SU(\infty)$ to smaller groups, in particular $SU(2)$ means that they can be easily break and separate from the pool of the $SU(\infty)$ symmetry—for instance by quantum correlation between pairs of subsystems—without affecting the infinite symmetry. And indeed it seems to be the case because $SU(2)$ and $SU(3) \subset SU(2) \times SU(2) \cong SU(4)$ are Standard Model symmetries. Additionally, string theory is fundamentally first quantized and string based field theories are considered to be low energy effective descriptions. However, as explained in the previous sections, in the present model owing to its infinite dimensional symmetry, Hilbert and Fock spaces are homomorphic and the model can be straightforwardly considered as first or second quantized.

The importance of $SU(2)$ symmetry in the construction of LQG and its presentation as spin foam [20] is shared with the present model. However, $SU(2) \cong SO(3)$ manifold on which Ashtekar variables are defined has its origin in the ADM (3 + 1)D formalism, based on the presumption that spacetime and thereby quantum gravity should be formulated in the physical spacetime. Moreover, LQG does not address the origin of matter as the source of gravity or the origin of the Standard Model symmetries. The present model explains both the dimension of spacetime and relation between quantum gravity, matter, and SM symmetries.

A concept that string theory and LQG does not consider—at least not in their foundation—is the fact that in quantum mechanics discrimination between observer and observered is essential, and models which do not consider this concept in their construction—especially when the models is intended to be applied to the whole Universe—are somehow metaphysical, because they implicitly consider that the observer is out of this Universe.

6. Outline and Future Perspectives

In this work, we proposed a new approach to quantum gravity by constructing a Universe in which gravity is fundamentally quantic and demonstrated how it may answer some of questions that we raised in the Introduction section regarding gravity and the nature of spacetime. As we have already summarized the model and its results in Section 1.1, here we concentrate on perspectives for further studies.

Understanding nonlocality and differential form of the algebra of subsystem defined by Equation (7) is crucial for finding an algebraic expression for the Lagrangian (13), which, at present, is too abstract. This task is especially important for investigating the semi-classical limit of the model. On the other hand, this Lagrangian describes an open system, because the state of the observer and probably some of degrees of freedom of the clock are traced out. Formulation of the subdivided Universe as an open system should help application of the model to black hole physics and cosmology.

We discussed a bottom-up procedure for the emergence of internal symmetries in Section 4. In particular, we concluded that they should generate stronger couplings between particles/subsystems than gravity. However, this argument does not explain how the hierarchy of couplings arises. We conjecture that clustering of subsystems, which leads to the emergence of internal symmetries, also determines their couplings, probably through processes that are analogous to the formation of moiré super-lattice and strong correlation between electrons in 2D materials. The fact that, in this model, both the Universe as a whole and its subsystems have $SU(\infty)$ symmetry, which is represented by diffeomorphism of 2D surfaces, means that the necessary ingredients for formation of moiré-like structures are readily available.

In the absence of experimental quantum gravity tests, the ability of models to solve theoretical issues has prominent importance. Among topics that must be addressed black holes and puzzles of information loss in semi-classical approaches have high priority. Because the model studied here is inherently quantic, the first task is finding a purely quantic definition for black holes. Naively, a quantum black hole may be defined as a many particle system in a quantum well in real space. However, we know that quantum field theory in curved spacetime background of black holes leads to Hawking radiation and extraction of energy from black hole. Consequently, in the realm of quantum mechanics, black holes are not really contained in a limited region of space. Their potential well is not perfect and their matter content extends to infinity. Thus, this definition should be considered as an initial condition.

Inflation and dark energy are other issues that should be investigated in the context of this model. Notably, it would be interesting to see whether the topological nature of 2D Lagrangian of the whole Universe can have observable consequences, for instance, as a small but nonzero vacuum energy.

As for inflation, an exponential decoupling and decoherence of particles/subsystems in the early universe may be interpreted as inflation and an extension of spacetime. These possibilities need detailed investigation.

In conclusion, the inhomogeneous Lorentz transformation may be the classical interface of a much deeper and global realm of a quantum Universe.

Funding: This research received no external funding.

Acknowledgments: The author thanks Institut Henry Poincaré for hospitality and bibliographic assistance during accomplishment of this work.

Conflicts of Interest: The author declares no conflict of interest.

Appendix A. A Very Brief Summary of the Best Studied Quantum Gravity Models

Introduction of quantum mechanical concepts to general relativity was first mentioned by Einstein himself in his famous 1916 paper. The first detailed work on the topic was by Léon Rosenfeld in 1930 [41], in which the action of Einstein-Hilbert model with matter is quantized by replacing classical variables with hermitian operators, see e.g., [42] for the history of early approaches to quantum gravity. This canonical approach and its modern variants based on the quantization of 3 + 1 dimensional Hamiltonian description of dynamics, notably Wheeler-DeWitt (WD) formalism [43,44] and quantum geometrodynamics [45] lead to nonrenormalizable models.[13] See e.g., [46] for review of other issues of these approaches and their current status.

Another model, inspired by the ADM Hamiltonian formulation of general relativity [47], the Dirac Hamiltonian description of quantum mechanics [48], and the WD approach to QGR is Loop Quantum Gravity (LQG), see e.g., [49,50] and references therein. In this approach, triads defined on a patch of the 3D space—what is called Ashtekar variables [19]—replace spatial coordinates and are considered as Hermitian operators acting on the Hilbert space of the Universe. Their conjugate operators form a $SU(2)$ Yang-Mills theory and provide a connection—up to an undefined constant called Immirzi parameter—for the quantized 3D space. However, to implement diffeomorphism of general relativity without referring to a fixed background, the physical quantized entities are holonomies—gauge invariant nonlocal fluxes and Wilson loops defined on 2D surfaces and their boundaries, respectively. Similar to the WD formalism, the LQG Hamiltonian is a constraint, and thereby there is no explicit time in the model [49]. Recently, it is shown that a conformal version of the LQG has an explicit time parameter [51]. But, conformal symmetry must be ultimately broken to induce a mass or distance scale in the model. Other issues in the LQG are lack of explicit global Lorentz invariance, absence of any direct connection to matter, and most importantly quantization of space, that violates Lorentz invariance even when the absence of time parameter in the model is neglected.

Regrading the violation of Lorentz invariance, even if discretization is restricted to distances close to the Planck scale, matter interaction propagates it to larger distances [10]. This issue is also present in other background independent approaches to quantum gravity, in which in one way or another the spacetime is discretized. Examples of such models are symplectic quantum geometry [52] and dynamical triangulations, in which space is assumed to consist of a dynamical lattice [53,54]. See also [55] for a recent review of these approaches and [56] for some of their issues, in particular a likely absence of a UV fixed point, which is necessary for renormalizabilty of these models. Therefore, the claimed quantization of space volume or in other words emergence of a fundamental length scale in UV limit of these models is still uncertain. Another example is causal sets—a discretization approach with causally ordered structures [57], see e.g., [58] for a review. They probably suffer from the same issue as other discretization models, notably breaking of Lorentz symmetry, see e.g., [59], but also [60]

[13] We should emphasize that references given in this appendix are only examples of works on the subjects on which tens or even hundreds of articles can be found in the literature.

for counter-arguments. We should remind that all quantum gravity models depend on a length (or equivalently mass) scale, namely the Planck length L_P (or mass M_P). Dimensionful quantities need a *unit*, which does not arise from dimensionless or scale invariant quantities. Therefore, discretization is not a replacement for a dimensionful fundamental constant in quantum gravity models.

Another way of quantizing spacetime without discretization is consideration of a noncommutative spacetime [61,62]. This formalism is in fact one of the earliest proposals for a quantum gravity. More recently this approach is studied in conjunction with other QGR models such as string theory [63] and matrix models [64]. An essential issue of this class of models is their inherent nonlocality that leads to mixing of low and high energy scales [65]. On the other hand, this characteristic might be useful for constraining them, and thereby related QGR models [66].

In early 1980's the discovery of both spin-1 and spin-2 fields in 2D conformal quantum field theories embedded in a D-dimensional spacetime—called *string* models—opened a new era and discipline for seeking a reliable quantum model for gravity, and ultimately unifying all fundamental forces in a *Great Unified Theory (GUT)*. Nambu-Goto and Polyakov string theories were studied in 1970's as candidates for describing strong interaction of hadrons. Although with the establishment of Quantum Chromo-Dynamics (QCD) as the true description of strong nuclear force string theories seemed irrelevant, their potential for quantizing spacetime [67,68] gave them a new role in fundamental particle physics. String and superstring theories became and continue to be by far the most extensively studied candidates of quantum gravity and GUT.[14]

Quantized strings/superstring models are finite and meaningful only for special values of spacetime dimension D. For these cases, the central charge of Virasoro algebra or its generalization to affine Lie algebra vanishes when the contribution of all fields, including ghosts of the conformal theory on the 2D world-sheet are taken into account. Without this restriction the theory is infested by anomalies, singularities, and misbehaviour. The allowed dimension is $D = 26$ for bosonic string theories and $D = 10$ for superstrings. The group manifold on which a viable string model can live is restricted as well. For instance, the allowed symmetry in heterotic Polyakov model is $SO(32)$ or $E_8 \times E_8$. Wess-Zumino-Novikov-Witten (WZNW) models with 2D affine Lie algebras provide more variety of symmetries, including coset groups. However, restriction on dimension/rank of symmetry groups remains the same. Therefore, to make contact with real world, which has 3 + 1 dimensions, the remaining dimensions must be compactified.

Initially the inevitable compactification of fields in string models was welcomed because it might explain internal global and local (gauge) symmetries of elementary particles, in a similar manner as in Kaluza-Klein unification of gravity and electromagnetism [71,72]. However, intensive investigations of the topic showed that compactification generates a plethora of possible models. Some of these models may be considered more realistic than others based on the criteria of having a low energy limit containing the Standard Model symmetries. But, unobserved massless moduli, which may make the Universe overdense if they acquire a mass at string or even lower scales, strongly constrain many of string models. Therefore, moduli must be stabilized [73,74]. For instance, they should acquire just enough effective mass to make them a good candidate for dark matter [75]. Moreover, in string theories there is no natural inflation candidate satisfying cosmological observations without fine-tuning. Although moduli are considered as potential candidates for inflation [76], small non-Gaussianity of Cosmic Microwave Background (CMB) anisotropies [77] seems to prefer single field inflation [78]. In addition, single field slow roll inflation may be inconsistent [79] with constraints to be imposed on a scalar field interacting with quantum gravity in the framework of swampland extension of string models landscape [80]. Some researchers still believe that a genuinely non-perturbative formulation of superstring theories may solve many of these issues [15]. However, the absence of any evidence of

[14] A textbook description and references to original works can be found in textbooks such as [69,70].

[15] Non-supersymmetric string models may have no non-perturbative formulation and should be considered as part of a supersymmetric model, see e.g., Chapter 8 of [70].

supersymmetry up to $\sim$TeV energies at LHC—where it was expected, such that it could solve Higgs hierarchy problem [81]—is another disappointing result for string models.

Observation of accelerating expansion of the Universe due to a mysterious dark energy with properties very similar to a cosmological constant—presumably a nonzero but very small vacuum energy—seems to be another big obstacle for string theory [82] as the only quantum gravity candidate including both matter and gravity in its construction. The landscape of string vacua has $\gtrsim 10^{200-500}$ minima—depending on how models are counted [83]. But there is no rule to determine which one is more likely and why the observed density of dark energy—if it is the vacuum energy—is $\sim 10^{123}$ fold less than its expected value, namely $M_P{}^4$. To tackle and solve some of these issues, extensions and/or reformulations of string theories have led to their variants such as matrix models [84,85], M-theory, F-theory, and more recently swampland [80] and weak gravity conjecture [38,86], and models constructed based on them.

In early 1999 Randall-Sundrum brane models [87,88] and their variants—inspired by D-branes in toroidal compactification of open strings and propagation of graviton closed strings in the bulk of one or two non-compactified warped extra dimensions—generated a great amount of excitement and were subject of intensive investigations. By confining all fields except gravitons on 4D branes these models are able to lower the fundamental scale of quantum gravity to TeV energies—presumably the scale of weak interaction—and explain the apparent weakness of gravitational coupling and high value of Planck mass. Thus, a priori brane models solve the problem of coupling hierarchy in Standard Model of particle physics. In addition, an effective small cosmological constant on the visible brane may be achievable [89,90]. However, brane models, in general, have a modified Friedmann equation, which is strongly constrained by observations [91–93]. Moreover, it is shown that the confinement of gauge bosons on the brane(s) violates gauge symmetries, and if gauge fields propagate to the bulk, so do the matter [94,95]. Nonetheless, some methods for their localization on the brane are suggested [96,97]. On the other hand, observation of ultra high energy cosmic rays constrains the scale of quantum gravity and characteristic scale of warped extra-dimension to >100 TeV [98,99]. This constraint is consistent with other theoretical and experimental issues of brane models, specially in the context of black hole physics, that is instability of macroscopic black holes, nonexistence of an asymptotically Minkowski solution [100,101], and observational constraint [102] on the formation of microscopic black holes in colliders at TeV energies [103].

In the view of these difficulties more drastic ideas have emerged. Some authors suggests UV/IR correspondence of gravity. They propose that at UV scales graviton quantum condensate behaves asymptotically similar to classical gravity [104,105]. Other proposals attracting some interest include the emergence of classical gravity and spacetime from thermodynamics and entropy [106,107] or condensation of more fundamental fields [108,109].

More recently, the development of quantum information theory and its close relation with entanglement of quantum systems, their entropy and the puzzle of information loss in Hawking radiation of black holes have promoted models that interpret gravity and spacetime as an emergent effect of entanglement [110–112] and tensor networks [113,114]. These ideas are in one way or another related to holography principle and Ads/CFT equivalence conjecture [115]. In these models spacetime metric and geometry emerge from tensor decomposition of the Hilbert space of the Universe to entangled subspaces. The resulted structures are interpreted as graphs and a symplectic geometry is associated to them. In the continuum limit the space of graphs can be considered as a quantum spacetime. In a somehow different approach in the same category of models the concept of *locality* specified by subalgebras is used to decompose the Universe. Local observables belong to spacelike subspaces in a given reference frame/basis [116,117]. This means that in these models a background spacetime is implicitly postulated without being precise about its origin and nature. In addition to spacetime, subsystems/subalgebras should somehow present matter. But, it is not clear how they are related. Moreover, the problem of the spacetime dimensionality and how it acquires its observed value is not discussed. In any case, investigation of these approaches to quantum gravity is still in its infancy and their theoretical and observational consistency are not fully worked out.

Appendix B. Quantum Mechanics Postulates in Symmetry Language

In this appendix we reformulate axioms of quantum mechanics à la Dirac [118] and von Neumann [119] with symmetry as a foundational concept:

i. A quantum system is defined by its symmetries. Its state is a vector belonging to a projective vector space called *state space* representing its symmetry group. Observables are associated to self-adjoint operators. The set of independent observables is isomorphic to subspace of commuting elements of the space of self-adjoint (Hermitian) operators acting on the state space and generates the maximal abelian subalgebra of the algebra associated to symmetry group.
ii. The state space of a composite system is homomorphic to the direct product of state spaces of its components.[16] In the special case of separable components, this homomorphism becomes an isomorphism. Components may be separable-untangled—in some symmetries and inseparable—entangled—in others. The symmetry group of the states of a composite system is a subgroup of direct product of its components.
iii. Evolution of a system is unitary and is ruled by conservation laws imposed by its symmetries and their representation by the state space.
iv. Decomposition coefficients of a state to eigen vectors [17] of an observable presents the coherence/degeneracy of the system with respect to its environment according to that observable. Projective measurements **by definition** correspond to complete breaking of coherence/degeneracy. The outcome of such measurements is the eigen value of the eigen state to which the symmetry is broken. This spontaneous decoherence (symmetry breaking) [18] reduces the state space to the subspace generated by other independent observables, which represent remaining symmetries/degeneracies.
v. A probability independent of measurement details is associated to eigen values of an observable as the outcome of a measurement. It presents the amount of coherence/degeneracy of the state before its breaking by a projective measurement. Physical processes that determine the probability of each outcome are collectively called *preparation*.[19]

These axioms are very similar to their analogues in the standard quantum mechanics, except that we do not assume an abstract Hilbert space. The Born rule and classification of the state space as a Hilbert space can be demonstrated using axioms (i) and (v), and properties of statistical distributions [21]. We remind that the symmetry represented by the Hilbert space of a quantum system is in addition to the global $U(1)$ symmetry of states, which leaves probability of outcomes in a projective measurement unchanged. When system is divided to subsystems that can be approximately considered as non-interacting, each subsystem acquire its own local $U(1)$ symmetry. Even in presence of interaction between subsystems, a local $U(1)$ symmetry can be considered, as long as the interaction does not change the Hilbert space of subsystems. We notice that axiom (ii) slightly diverges from its analogue in the standard quantum mechanics. It emphasizes on the fact that the symmetry group represented by a composite system can be smaller than the tensor product of those of its components.

16 Notice that this axiom differentiates between possible states of a composite system, which is the direct product of those of subsystems, and what is actually realized, which can be limited to a subspace of the direct product of individual components and have reduced symmetry.
17 More precisely *rays* because state vectors differing by a constant are equivalent.
18 Ref. [21] explains why decoherence should be considered as a spontaneous symmetry breaking similar to a phase transition.
19 Literature on the foundation of quantum mechanics consider an intermediate step called *transition* between preparation and measurement. Here we include this step to preparation or measurement operations and do not consider it as a separate physical operation.

In particular, entanglement may reduce the dimension of Hilbert space and thereby the rank of symmetry group that it represents.

A corollary of these axioms is that without division of the Universe to *system(s)* and *observer(s)* the process of measurement is meaningless. In another word, an indivisible universe is trivial and homomorphic to an empty set. In standard quantum mechanics the necessity of the division of the Universe to subsystems arises in the Copenhagen interpretation, which has many issues, see e.g., [120] for a review. In covariant quantum models and ADM canonical quantization of gravity, in which Hamiltonian is always null and naively the Universe seems to be static, relational definitions of time is based on the division of the Universe to subsystems, see e.g., [27]. Therefore, we conclude that division to subsystems is fundamental concept and must be explicitly included in the construction of quantum cosmology models.

Appendix C. State Space Symmetry and Coherence

The choice of a Hilbert space $\mathcal{H}$ to present possible states of a system is usually based on the symmetries of its classical Lagrangian. Although these symmetries have usually a finite rank—the number of simultaneously measurable observables—the Hilbert space presenting them may be infinite dimensional. For example, translation symmetry in a 3D space is homomorphic to $U(1) \times U(1) \times U(1)$ and has a global $SU(2) \cong SO(3)$ symmetry under rotation of coordinates. They can be presented by 6 parameters/observables. Thus, the rank of the symmetry is finite. Nonetheless, due to the abelian nature of $U(1)$ group, the Hilbert space of position operator $\mathcal{H}_X$ is infinite dimensional. More generally, the dimension of the Hilbert space depends on the dimension of the representation of the symmetry group of Lagrangian and its reducibility. The Hilbert space of a multi-particle system can be considered as a reducible representation of the symmetry, even if single particles are in an irreducible representation. In particular, Fock space of a many-particle system can be presented as an infinite dimensional Hilbert space representing symmetries of the Lagrangian in a reducible manner. This property is important for the construction of the quantum Universe model studied here, because it demonstrates that the infinite size of the physical space can be equally interpreted as manifestation of infinite number of particles/subsystems in a composite Universe.

Ensemble of linear operators acting on a Hilbert space $\mathcal{B}[\mathcal{H}]$ represents $SU(N)$ group where N is the dimension of the Hilbert space $\mathcal{H}$ and can be infinite. As discussed in details in [21] configuration space of classical (statistical) systems have $\bigotimes^N U(1)$ symmetry where each $U(1)$ is isomorphic to the continuous range of values that an observable may have. Thus, quantization extends the symmetry of classical configuration space to $\mathcal{B}[\mathcal{H}] = SU(N)xU(1) \cong U(N) \supset \bigotimes^N U(1)$, where here we have also considered the global $U(1)$ symmetry of the Hilbert space.

Application of linear operators can be interpreted as interaction with another system or more generally with the rest of the Universe. The change of state can be also considered as Positive Operator Valued Measurement (POVM). In particular, a projective measurement and decoherence makes the state completely incoherent $\hat{\rho}_{inc}$:

$$\hat{B}\hat{\rho}_c\hat{B}^\dagger \rightarrow \hat{\rho}_{inc} = \sum_i \rho_i \hat{\rho}_i, \qquad \hat{\rho}_i \equiv |i\rangle\langle i| \tag{A1}$$

where $\hat{B} \in \mathcal{B}[\mathcal{H}]$, $|i\rangle$ is an eigen basis for the measured observable, and subscript $_{inc}$ means incoherent. We remind that the space of simultaneously observable operators corresponds to the Cartan subalgebra of $\mathcal{B}[\mathcal{H}]$. Coefficients ρ_i are probability of occurrence of eigen value of $|i\rangle$ as outcome of the measurement. Because $\hat{\rho}_{inc}$ is diagonal, completely incoherent states $\hat{\rho}_{inc}$ represent the Cartan subgroup of $\mathcal{B}[\mathcal{H}]$. A maximally coherent state in the above basis is defined as:

$$\hat{\rho}_{maxc} \propto \sum_{i,j} |i\rangle\langle j| \tag{A2}$$

This is a pure state, in which all eigen states have the same occurrence probability in a projective measurement. Notice that due to the projectivity of Hilbert space $\hat{\rho}_{maxc}$ is unique and application of any other member of $\mathcal{B}[\mathcal{H}]$ reduces its coherence, quantified for instance by fidelity or Fubini-Study metric [22]. More generally, action of $\mathcal{B}[\mathcal{H}]$ members changes coherence of any state which is not completely incoherent. For this reason, we call $SU(N)$ symmetry of $\mathcal{B}[\mathcal{H}]$ the *coherence symmetry*.[20]

It is useful to remind that in particle physics generators of $\mathcal{B}[\mathcal{H}]$ space physically exist and are not abstract operation of an apparatus controlled by an experimenter. In the Standard Model $\mathcal{B}[\mathcal{H}]$ is generated by vector boson gauge fields in fundamental representation of SM symmetry group. They act on the Hilbert space generated by matter fields. If gravity, which is the only known universal interaction follows the same rule, we should be able to define a Hilbert space for matter on which linear operators representing gravity act. In the model studied here we identify these operators with $\hat{L}$'s defined in Section 2.

Regarding the example of translational and rotational symmetries of the physical space mentioned earlier, despite the fact that the dimension of the Cartan subalgebra of $\mathcal{B}[\mathcal{H}_X] \cong SU(N \to \infty)$ is infinite, and a priori there must be infinite simultaneously observable quantities in the physical space, in quantum mechanics only one vector observable is associated to $\mathcal{B}[\mathcal{H}_X]$, namely the position of a particle/system. QFTs define field operators at every point of the space and assume that at equal time operators at different positions commute (or in the case of fermions anti-commute). However, in the formulation of QFT models position is a parameter not an operator. These different interpretations of spacetime highlight the ambiguity of its nature in quantum contexts—as described in question 2 in the Introduction section.

Appendix D. $SU(\infty)$ and Its Polynomial Representation

Special unitary group $SU(N)$ can be considered as N-dimensional representation of $SU(2)$. For this reason generators $T^{(N)}_{lm}$ of the associated Lie algebra $su(N)$ can be expanded as a matrix polynomial of N-dimensional generators of $SU(2)$. Indices (l, m) in these generators are the same as in $SU(2)$ representations: $l = 0, \cdots, N-1$, $m = -l \ldots, +l$. Lie bracket of generators $T^{(N)}_{lm}$ is defined as:

$$[\hat{T}^{(N)}_{lm}, \hat{T}^{(N)}_{l'm'}] = f^{(N)\, l''m''}_{lm,l'm'} \hat{T}^{(N)}_{l''m''} \tag{A3}$$

Structure coefficients $f^{(N)\, l''m''}_{lm,l'm'}$ of $su(N)$ can be written with respect to 3j and 6j symbols, see e.g., [29] for their explicit expression. For $N \to \infty$, after rescaling these generators $\hat{T}^{(N)}_{lm} \to (N/i)^{1/2}\hat{T}^{(N)}_{lm}$, they satisfy the following Lie brackets:

$$[\hat{L}_{lm}, \hat{L}_{l'm'}] = f^{l''m''}_{lm,l'm'} \hat{L}_{l''m''} \tag{A4}$$

where $\hat{L}_{lm} \equiv \hat{T}^{(N\to\infty)}_{lm}$ and coefficients $f^{l''m''}_{lm,l'm'}$ are $N \to \infty$ limit of $f^{(N)\, l''m''}_{lm,l'm'}$. In addition, it is shown [29] that $\hat{L}_{lm}$ can be expanded with respect to spherical harmonic functions $Y_{lm}(\theta, \phi)$ defined on a sphere, i.e., the manifold associated to $SU(2)$:

$$\hat{L}_{lm} = \frac{\partial Y_{lm}}{\partial \cos\theta} \frac{\partial}{\partial \phi} - \frac{\partial Y_{lm}}{\partial \phi} \frac{\partial}{\partial \cos\theta} \tag{A5}$$

$$\hat{L}_{lm} Y_{l'm'} = -\{Y_{lm}, Y_{l'm'}\} = -f^{l''m''}_{lm,l'm'} Y_{l''m''} \tag{A6}$$

$$\{\mathrm{f}, \mathrm{g}\} \equiv \frac{\partial \mathrm{f}}{\partial \cos\theta} \frac{\partial \mathrm{g}}{\partial \phi} - \frac{\partial \mathrm{f}}{\partial \phi} \frac{\partial \mathrm{g}}{\partial \cos\theta}, \quad \forall\, \mathrm{f}, \mathrm{g} \text{ defined on the sphere} \tag{A7}$$

---

[20] In some quantum information literature coherence symmetry is called *asymmetry* [23]. In this work we call it *coherence symmetry* or simply *coherence* to remind that its origin is quantum degeneracy and indistinguishability/symmetry of states before a projective observation.

where $\theta = [0, \pi]$ and $\phi = [0, 2\pi]$ are angular coordinates and $\{\mathrm{f},\ \mathrm{g}\}$ is the Poisson bracket of continuous functions f and g on the sphere. We notice that although generators $\hat{L}_{lm}$ are linear combination of $\partial/\partial(\cos\theta)$ and $\partial/\partial\phi$, the latter operators cannot be considered as generators of $SU(N \to \infty)$, because they commute with each other and generate only the abelian subspace of $SU(\infty)$ group.

Using (A5)–(A7), coefficients $f_{lm,l'm'}^{l''m''}$ can be determined:

$$f_{lm,l'm'}^{l''m''} = \frac{(2l''+1)}{4\pi} \int d^2\Omega\, Y^*_{l''m''}\{Y_{lm},\ Y_{l'm'}\}, \qquad Y^*_{lm} = Y_{l,-m} \qquad d^2\Omega \equiv d(\cos\theta)\, d\phi \tag{A8}$$

Here we normalize Y_{lm} such that:

$$\int d^2\Omega\, Y^*_{l'm'}\, Y_{lm} = \frac{4\pi}{(2l+1)} \delta_{ll'}\, \delta_{mm'} \tag{A9}$$

Although $\hat{L}_{lm}$ is defined in discrete (l, m) space—analogous to a discrete Fourier mode—we can use inverse expansion to define operators which depend only on continuous angular coordinates:

$$\hat{L}(\theta, \phi) \equiv \sum_{l,m} Y^*_{lm} \hat{L}_{lm} \tag{A10}$$

As $\{\hat{L}(\theta, \phi)\}$ are linear in $\hat{L}_{lm}$ and contain all these generators, they are also generators of $SU(N \to \infty) \cong \mathrm{Diff}(S_2)$ and coefficients in their Lie bracket is expressed with respect to θ and ϕ as:

$$\mathtt{f}((\theta,\phi),(\theta',\phi');(\theta'',\phi'')) = \sum_{lm,l'm',l''m''} Y^*_{lm}(\theta,\phi) Y^*_{l'm'}(\theta',\phi') Y_{l''m''}(\theta'',\phi'') f_{lm,l'm'}^{l''m''} \tag{A11}$$

Coefficients f are anti-symmetric with respect to the first two sets of parameters and can be considered as a 2-form on the sphere, and Lie algebra of $\hat{L}(\theta, \phi)$ operators as:

$$[\hat{L}(\theta_1,\phi_1), \hat{L}(\theta_2,\phi_2)] = \int d\Omega_3\, \mathtt{f}((\theta_1,\phi_1),(\theta_2,\phi_2);(\theta_3,\phi_3))\, \hat{L}(\theta_3,\phi_3) \tag{A12}$$

Operators $\hat{L}(\theta, \phi)$ are continuous limits of $\hat{L}_{lm}$'s and both set of generators are vectors and live on the tangent space of the sphere.

Appendix E. Cartan Decomposition of $SU(\infty)$

Representations of $su(N)$ algebra can be decomposed to direct sum of smaller su algebras, see e.g., [34] and references therein. In the case of $SU(\infty)$ the fact that its algebra is homomorphic to Poisson brackets of spherical harmonic functions, which in turn correspond to representations of $SU(2) \cong SO(3)$, means that $su(\infty)$ algebra should be expandable as direct sum of representations of $SU(2)$, see e.g., [29,30] for the proof. Thus, up to a normalization factor depending only on l, generators of $su(\infty)$ algebra $\hat{L}_{lm}$ can be expanded as:

$$\hat{L}_{lm} = \mathcal{R} \sum_{i_\alpha=1,\,2,\,3, \alpha=1,\cdots,l} a^{(m)}_{i_1,\cdots i_l} \sigma_{i_1} \cdots \sigma_{i_l} \tag{A13}$$

where σ_{i_α}'s are $N \to \infty$ representation of Pauli matrices [29]. Coefficients $a^{(m)}$ are determined from expansion of spherical harmonic functions with respect to spherical description of Cartesian coordinates, see [29] for details. This explicit description shows that up to a constant factor $\hat{L}_{lm}$ operators can be considered as tensor product of 2×2 Pauli matrices, and $SU(\infty) \cong SU(2) \otimes S(2) \otimes \ldots$. This relation can be understood from properties of $SU(N)$ group. Specifically, $SU(N) \supseteq SU(N-K) \otimes SU(K)$. For $N \to \infty$ and finite K, $SU(N - K \to \infty) \cong SU(\infty)$. Therefore $SU(\infty)$ is homomorphic to infinite tensor product of SU groups of finite rank, in particular $SU(2)$—the smallest non-abelian SU group. This shows that $SU(2)$ group, which has a key role in some quantum gravity models, notably in LQG, simply presents a mathematical description rather than a fundamental physical entity. The description of $SU(\infty)$ as tensor product of $SU(2)$ is comparable with Fourier transform, which presents the simplest decomposition to orthogonal functions, but can be replaced by another orthogonal function. It is only the application that determines which one is more suitable.

Appendix E.1. Eigen Functions of $\hat{L}(\theta,\phi)$ and $\hat{L}_{lm}$

We define eigen functions of $\hat{L}(\theta,\phi)$ and $\hat{L}_{lm}$ operators as the followings:

$$\hat{L}(\theta,\phi)\eta(\theta,\phi) = N\eta(\theta,\phi) \tag{A14}$$

$$\hat{L}_{lm}\zeta_{lm} = N'\zeta_{lm} \tag{A15}$$

where N and N' are constants[21], but N' may depend on (l,m). Using definition of $\hat{L}(\theta,\phi)$ and $\hat{L}_{lm}$ and properties of spherical harmonic functions, solutions of Equations (A14) and (A15) are obtained as:

$$\begin{cases} \eta(\theta,\phi) = iN\sum\limits_{lm} \frac{(l+m)!}{mA_l(l-m)!}\left[\mathcal{F}_{lm}(\cos\theta) - \mathcal{F}_{lm}(\cos\theta_0(t))\right] + \eta(\theta_0(t)) \\ \phi + H(\cos\theta) = -[H(\cos\theta_0(t)) - \phi_0(t)] \end{cases} \tag{A16}$$

$$A_l \equiv \sqrt{\frac{4\pi}{2l+1}} \qquad \mathcal{F}_{lm} \equiv \int d(\cos\theta)|P_{lm}(\cos\theta)|^{-2}, \tag{A17}$$

$$H(\cos\theta) \equiv \int d(\cos\theta)\frac{\sum\limits_{lm}\frac{A_l(l-m)!}{(l+m)!}\frac{\partial|P_{lm}\cos\theta|^2}{2\partial\cos\theta}}{\sum\limits_{l'm'}\frac{im'A_{l'}(l'-m')!}{(l'+m')!}|P_{l'm'}(\cos\theta)|^2} \tag{A18}$$

where t parameterizes tangent surface at initial point (θ_0,ϕ_0). Elimination of this parameter from two equations in (A16) determines $\eta(\theta,\phi)$ for a set of initial conditions. Because the second equation does not depend on N, without loss of generality we can scale initial value $\eta(\theta_0) \to iN\eta(\theta_0)$. With this choice the eigen value N can be factorized, and because Hilbert space is projective, N can be considered as an overall normalization factor and irrelevant for physics. Therefore, each set of parameters (θ,ϕ) present a unique pointer state for the Hilbert space.

In the same way we can calculate eigen functions of $\hat{L}_{lm}$ as a parametric function:

$$\begin{cases} \zeta_{lm}(\theta) = -N'e^{m^2W_{lm}(\theta_0)}\sqrt{\frac{(l+m)!}{(l-m)!}}\left[Z_{lm}(\theta) - Z_{lm}(\theta_0(t))\right] + \zeta_{lm}(\theta_0(t)) \\ \phi - imW_{lm}(\theta) = \phi_0(t) - imW_{lm}(\theta_0(t)) \end{cases} \tag{A19}$$

$$W_{lm} \equiv \int d(\cos\theta)\frac{(1-\cos\theta^2)\,P_{lm}(\cos\theta)}{(l-m+1)\,P_{(l+1)m}(\cos\theta)} - (l+1)P_{lm}(\cos\theta) \tag{A20}$$

$$Z_{lm} \equiv \int d(\cos\theta)\frac{e^{-m^2\,W_{lm}(\cos\theta)}}{(l-m+1)\,P_{(l+1)m}(\cos\theta)} - (l+1)P_{lm}(\cos\theta) \tag{A21}$$

Similar to $\eta(\theta,\phi)$, redefinition of initial value $Z_{lm}(\theta_0(t)) \to Z_{lm}(\theta_0(t))N'e^{m^2W_{lm}(\theta_0)}$ leads to a unique eigen function for $\hat{L}_{lm}$.

[21] A priori N and N' can depend on (θ,ϕ). However, their dependence on angular parameters can be included in η. Therefore, only constant eigen values matter.

Considering diffeomorphism invariance of the model, it is always possible to redefine coordinates such that $\theta = const.$ and $\phi = const.$ constitute a basis and any state can be written as:

$$|\psi\rangle = \int d^2\Omega \, \psi(\theta,\phi)|\theta,\phi\rangle \tag{A22}$$

Thus, as explained in the main text, vectors of the Hilbert space representing $SU(\infty)$ are complex functions on 2D surfaces. As $SU(\infty) \cong \text{Diff}(S_2)$, the range of (θ,ϕ) is $\theta = [0.,\pi]$ and $\phi = [0.,2\pi)$. However, $SU(\infty)$ may be represented by diffeo-surfaces of higher genus. In this case $|\theta + n\pi, \phi + 2n'\pi\rangle$ for any integer n and n' may present different states. States can be also expanded with respect to $|l,m\rangle$ [29].

Appendix E.2. Dynamics Equations of the Universe before its Division to Subsystems

The equilibrium solution for Lagrangian $\mathcal{L}_U$ in (3) can be determined by variation with respect to L_{lm} and components of the state ρ in an orthogonal basis of the Hilbert space. In absence of environment for the whole Universe, ρ is pure and can be written as $\rho = |\psi\rangle\langle\psi|$, where $\psi\rangle$ is an arbitrary vector in the Hilbert space $\mathcal{H}_U$. As discussed in Section 2 and explicitly shown in Appendix E.1, vectors of $\mathcal{H}_U$ correspond to complex functions of angular coordinates (θ,ϕ) of the diffeo-surface, and can be expanded with respect to spherical harmonic functions. Nonetheless, here we follow the usual bracket notation of quantum mechanics and call states of this orthogonal basis $|l,m\rangle$, where $l \in \mathbb{Z}/2$, $-l \leqslant m \leqslant +l$. In this basis $|\psi\rangle = \sum_{l,m} \psi_{lm} \, |l,m\rangle$ and $\rho = \sum_{l,m,l',m'} \psi_{l,m}\psi^*_{l',m'}|l,m\rangle\langle l',m'|$. After this decomposition dynamics equations are expressed as:

$$\frac{\partial \mathcal{L}_U}{\partial \psi_{lm}} = \sum_{l',m',l'',m''} L_{l''m''}\psi^*_{l'm'}\langle l'm'|\hat{L}_{l''m''}|l,m\rangle \tag{A23}$$

$$\frac{\partial \mathcal{L}_U}{\partial L_{lm}} = \sum_{l',m',l'',m''} \psi^*_{l''m''}\psi_{l''m''}\langle l',m'|\hat{L}_{lm}|l''m''\rangle + 2L_{lm}\text{tr}(\hat{L}_{lm}\hat{L}_{lm}) \tag{A24}$$

Because $\hat{L}_{lm}$ is a generator of $SU(\infty)$, the last term in (A24) is a constant depending only on l and normalization of generators. Thus, we define $C_l \equiv \text{tr}(\hat{L}_{lm}\hat{L}_{lm})$. Using description of $\hat{L}_{lm}$ in (A13) to tensor product of Pauli matrices, we conclude that $\hat{L}_{lm}|l',m'\rangle \neq 0$ only for $l \geqslant l'$ and consists of linear combination of $|l'',m''\rangle$ states. On the other hand, $\langle l',m'|l,m\rangle = \delta_{ll'}\delta_{mm'}$. Thus, $\langle l',m'|\hat{L}_{lm}|l''m''\rangle$ is nonzero only for terms with equal l indices and we can solve (A24) for L_{lm} as the following:

$$L_{lm} = -\frac{1}{2C_l} \sum_{|m'|,|m''| \leqslant l, m+m'+m''=0} \psi^*_{lm'}\psi_{lm''}\langle l,m'|\hat{L}_{lm}|lm''\rangle \tag{A25}$$

By applying this solution to (A23) and using properties of $\hat{L}_{lm}$ and $|l,m\rangle$ we find:

$$\sum_{m'',m''} \psi^*_{l,-(m+m')}\psi^*_{l,-(m+m'')}\psi_{lm'}\langle l,-(m+m'')|\hat{L}_{lm''}|l,m\rangle\langle l,-(m'+m'')|\hat{L}_{lm''}|l,m'\rangle = 0.$$
$$|m'|,|m'|,|m+m''|,|m'+m''|, \leqslant l, \forall l \in \mathbb{Z}/2 \tag{A26}$$

Considering independence and orthogonality of $|l,m\rangle$ states, this equation is satisfied only if $\psi_{lm} = 0$, $|m| \leqslant l$, $\forall l$. Thus, equilibrium solution of the Lagrangian $\mathcal{L}_U$ is a trivial vacuum.

References

1. Eppley, K.; Hanna, E. The Necessity of Quantizing the Gravitational Field. *Found. Phys.* **1977**, *7*, 51. [CrossRef]
2. Ziaeepour, H. And what if gravity is intrinsically quantic? *J. Phys. Conf. Ser.* **2009**, *174*, 012027 arXiv:0901.4634.

3. Barrow, J.D.; Magueijo, J. A contextual Planck parameter and the classical limit in quantum cosmology. *arXiv* **2020**, arXiv:2006.16036.
4. Kolb, E.W.; Turner, M.S. *The Early Universe*; Addison-Wesley Publishing Company: Boston, MA, USA, 1991.
5. Utiyama, R. Invariant theoretical interpretation of interaction. *Phys. Rev.* **1956**, *101*, 1597. [CrossRef]
6. Kibble, T.W.B. Lorentz invariance and the gravitational field. *J. Math. Phys.* **1961**, *2*, 212. [CrossRef]
7. Coleman, S.; Mandula, J. All possible symmetries of the *S* matrix. *Phys. Rev.* **1967**, *159*, 1251. [CrossRef]
8. Haag, R.; Lopuszanski, J.T.; Sohnius, M. All possible generators of supersymmetries of the *S* matrix. *Nucl. Phys. B* **1975**, *88*, 257. [CrossRef]
9. Percacci, R. Mixing internal and spacetime transformations: Some examples and counterexamples. *J. Phys. A* **2008**, *41*, 335403. [CrossRef]
10. Collins, J.; Perez, A.; Sudarsky, D.; Urrutia, L.; Vucetich, H. Lorentz invariance and quantum gravity: An additional fine-tuning problem? *Phys. Rev. Lett.* **2004**, *93*, 191301. [CrossRef]
11. Asenbaum, P.; Overstreet, C.; Kim, M.; Curti, J.; Kasevich, M.A. Atom-Interferometric test of the equivalence principle at the 10^{12} level. *arXiv* **2020**, arXiv:2005.11624.
12. Abdo, A.A.; Ackermann, M.; Ajello, M.; Asano, K.; Atwood, W.B.; Axelsson, M.; Baldini, L.; Ballet, J.; Barbiellini, G.; Baring, M.G.; et al. A limit on the variation of the speed of light arising from quantum gravity effects. *Nature* **2009**, *462*, 331. [CrossRef] [PubMed]
13. Wilczek, F. Riemann-Einstein Structure from Volume and Gauge Symmetry. *Phys. Rev. Lett.* **1998**, *80*, 4851. [CrossRef]
14. Torres-Gomez, A.; Krasnov, K. Gravity-Yang-Mills-Higgs unification by enlarging the gauge group. *Phys. Rev. D* **2010**, *81*, 085003. [CrossRef]
15. Barrett, J.W.; Kerr, S. Gauge gravity and discrete quantum models. *arXiv* **2013**, arXiv:1309.1660.
16. Birrell, N.D.; Davies, P.C.W. *Quantum Fields in Curved Space*; Cambridge University Press: Cambridge, UK, 1982.
17. Parker, L.; Toms, D. *Quantum Field Theory in Curved Spacetime*; Cambridge University Press: Cambridge, UK, 2009.
18. Ziaeepour, H. Issues with vacuum energy as the origin of dark energy. *Mod. Phys. Lett. A* **2012**, *27*, 1250154. [CrossRef]
19. Ashtekar, A. New Variables for Classical and Quantum Gravity. *Phys. Rev. Lett.* **1986**, *57*, 2244. [CrossRef]
20. Livine, E.R. Projected Spin Networks for Lorentz connection: Linking Spin Foams and Loop Gravity. *Class. Quant. Grav.* **2002**, *19*, 5525. [CrossRef]
21. Ziaeepour, H. Symmetry as a foundational concept in Quantum Mechanics. *J. Phys. Conf. Ser.* **2015**, *626*, 012074. [CrossRef]
22. Baumgratz, T.; Cramer, M.; Plenio, M.B. Quantifying Coherence. *Phys. Rev. Lett.* **2014**, *113*, 140401. [CrossRef]
23. Mondal, D.; Datta, C.; Sazim, S. Quantum Coherence Sets The Quantum Speed Limit For Mixed States. *Phys. Lett. A* **2016**, *380*, 689. [CrossRef]
24. Zanardi, P.; Lidar, D.; Lloyd, S. Quantum tensor product structures are observable-induced. *Phys. Rev. Lett.* **2004**, *92*, 060402. [CrossRef]
25. Mazenc, E.A.; Ranard, D. Target Space Entanglement Entropy. *arXiv* **2019**, arXiv:1910.07449.
26. Page, D.N.; Wootters, W.K. Evolution without evolution: Dynamics described by stationary observables. *Phys. Rev. D* **1983**, *27*, 2885. [CrossRef]
27. Hoehn, P.A.; Smith, A.R.H.; Lock, M.P.E. The Trinity of Relational Quantum Dynamics. *arXiv* **2019**, arXiv:1912.00033.
28. Cao, C.; Carroll, S.M.; Michalakis, S. Space from Hilbert Space: Recovering Geometry from Bulk Entanglement. *Phys. Rev. D* **2017**, *95*, 024031. [CrossRef]
29. Hoppe, J. Quantum Theory of a Massless Relativistic Surface and a Two-dimensional Bound State Problem. Ph.D. Thesis, MIT, Cambridge, MA, USA, 1982.
30. Floratos, E.G.; Iliopoulos, J.; Tiktopoulos, G. A note on $SU(\infty)$ classical Yang-Mills theories. *Phys. Lett. B* **1989**, *217*, 285. [CrossRef]
31. Hoppe, J. Diffeomorphism, Group, Quantization, and $SU(\infty)$. *Int. J. Mod. Phys. A* **1989**, *4*, 5235. [CrossRef]
32. Hoppe, J.; Schaller, P. Infinitely Many Versions of $SU(\infty)$. *Phys. Lett. B* **1990**, *237*, 407. [CrossRef]
33. Zunger, Y. Why Matrix theory works for oddly shaped membranes. *Phys. Rev. D* **2001**, *64* , 086003. [CrossRef]

34. Su, Z-Y. A Scheme of Cartan Decomposition for su(N). *arXiv* **2006**, arXiv:quant-ph/0603190.
35. Jacobson, T. Thermodynamics of Spacetime: The Einstein Equation of State. *Phys. Rev. Lett.* **1995**, *75*, 1260. [CrossRef]
36. Kitaev, A. Anyons in an exactly solved model and beyond. *Ann. Phys.* **2006**, *321*, 2. [CrossRef]
37. Nakamura, J.; Liang, S.; Gardner, G.C.; Manfra, M.J. Direct observation of anyonic braiding statistics at the $\nu = 1/3$ fractional quantum Hall state. *arXiv* **2006**, arXiv:2006.14115.
38. Arkani-Hamed, N.; Motl, L.; Nicolis, A.; Vafa, C. The String Landscape, Black Holes and Gravity as the Weakest Force. *J. High Energy Phys.* **2007**, *06*, 060. [CrossRef]
39. Mandelstam, L.; Tamm, I. The Uncertainty Relation Between Energy and Time in Non-relativistic Quantum Mechanics. *J. Phys. (USSR)* **1945**, *9*, 249.
40. Oppenheim, J.; Wehner, S. The Uncertainty Principle Determines Nonlocality of Quantum Mechanics. *Science* **2010**, *330*, 1072. [CrossRef]
41. Rosenfeld, L. Zur Quantelung der Wellenfelder. *Annal der Physik* **1930**, *397*, 113. [CrossRef]
42. Rocci, A. On first attempts to reconcile quantum principles with gravity. *J. Phys. Conf. Ser.* **2013**, *470*, 012004. [CrossRef]
43. Dewitt, B. Quantum Theory of Gravity. I. The Canonical Theory. *Phys. Rev.* **1967**, *160*, 1113. [CrossRef]
44. Hartle, J.B.; Hawking, S.W. Wave function of the Universe. *Phys. Rev. D* **1983**, *28*, 2960. [CrossRef]
45. Wheeler, J.A. On the nature of quantum geometrodynamics. *Ann. Phys.* **1957**, *2*, 604. [CrossRef]
46. Kiefer, C. Quantum geometrodynamics: Whence, whither? *Gen. Rel. Grav* **2009**, *41*, 877. [CrossRef]
47. Arnowitt, R.; Deser, S.; Misner, C. Dynamical Structure and Definition of Energy in General Relativity. *Phys. Rev.* **1959**, *116*, 1322. [CrossRef]
48. Dirac, P. Generalized Hamiltonian dynamics. *Proc. Roy. Soc. Lond. A* **1958**, *246*, 326. [CrossRef]
49. Rovelli, C. *Quantum Gravity*; Cambridge University Press: Cambridge, UK, 2004.
50. Ashtekar, A.; Lewandowski, J. Background Independent Quantum Gravity: A Status Report. *Class. Quant. Grav.* **2004**, *21*, R53. [CrossRef]
51. Wang, C.H.; Rodrigues, D.P.F. Closing the gaps in quantum space and time: Conformally augmented gauge structure of gravitation. *Phys. Rev. D* **2018**, *98*, 124041. [CrossRef]
52. Crnkovic, C. Symplectic geometry of covariant phase space. *Class. Quant. Grav.* **1988**, *5*, 1557. [CrossRef]
53. Migdal, A.A. Quantum Gravity as Dynamical Triangulation. In *Two Dimensional Quantum Gravity and Random Surfaces*; Gross, D.J., Piran, T., Weinberg, S., Eds.; World Scientific: Singapore, 1991; pp. 41–79.
54. Ambjorn, J.; Carfora, M.; Marzuoli, A. *The Geometry of Dynamical Triangulations*; Springer: Berlin/Heidelberg, Germany, 1997.
55. Ambjørn, J.; Görlich, A.; Jurkiewicz, J.; Loll, R. Quantum Gravity via Causal Dynamical Triangulations. In *Springer Handbook of Spacetime*; Ashtekar, A., Petkov, V., Eds.; Springer Handbooks; Springer: Berlin/Heidelberg, Germany, 2014.
56. Ambjorn, J.; Gizbert-Studnicki, J.; Goerlich, A.; Jurkiewicz, J.; Loll, R. Renormalization in quantum theories of geometry. *arXiv* **2002**, arXiv:2002.01693.
57. Sorkin, R.D. A Unitary Substitute for Continuous Topology. *Int. J. Theor. Phys.* **1991**, *30*, 923. [CrossRef]
58. Dowker, F. Causal sets and the deep structure of spacetime. In *100 Years of Relativity: Space-Time Structure-Einstein and Beyond*; Ashtekar, A., Ed.; World Scientific: Singapore, 2005; p. 445.
59. Mattingly, D. Causal sets and conservation laws in tests of Lorentz symmetry. *Phys. Rev. D* **2008**, *77*, 125021. [CrossRef]
60. Dowker, F.; Sorkin, R.D. Symmetry-breaking and zero-one law. *Class. Quant. Grav.* **2020**, *37*, 15. [CrossRef]
61. Snyder, H.S. Quantized Space-Time. *Phys. Rev.* **1947**, *71*, 38. [CrossRef]
62. Yang, C.N. On quantized space-time. *Phys. Rev.* **1947**, *72*, 874. [CrossRef]
63. Seiberg, N.; Witten, E. String Theory and Noncommutative Geometry. *J. High Energy Phys.* **1999**, *9909*, 032. [CrossRef]
64. Steinacker, H. Emergent Gravity from Noncommutative Gauge Theory. *J. High Energy Phys.* **2007**, *0712*, 049. [CrossRef]
65. Amelino-Camelia, G.; Mandanici, G.; Yoshida, K. On the IR/UV mixing and experimental limits on the parameters of canonical noncommutative spacetimes. *J. High Energy Phys.* **2004**, *0401*, 037. [CrossRef]
66. Hogan, C. Quantum Geometry and Interferometry. In *Proceedings of the 9th LISA Symposium*; Binétruy, P., Plagnol, E., Eds.; Astronomical Society of the Pacific: San Francisco, CA, USA, 2013.

67. Polyakov, A.M. Quantum geometry of bosonic string. *Phys. Lett. B* **1981**, *103*, 207. [CrossRef]
68. Polyakov, A.M. Quantum geometry of fermionic string. *Phys. Lett. B* **1981**, *103*, 211. [CrossRef]
69. Green, M.B.; Schwarz, J.H.; Witten, E. *Superstring Theory I & II*; Cambridge University Press: Cambridge, UK, 1987.
70. Polchinski, J. *String Theory I & II*; Cambridge University Press: Cambridge, UK, 2005.
71. , Kaluza, T. Zum Unitätsproblem der Physik. *Sitz. Preuss. Akad. Wiss. Phys. Math.* **1921**, *K1*, 966.
72. Klein, O. Quantentheori und fünfdimensionale Relativistätstheori. *Zeits Phys.* **1926**, *37*, 895. [CrossRef]
73. Nakamura, S.; Yamaguchi, M. Gravitino Production from Heavy Moduli Decay and Cosmological Moduli Problem Revived. *Phys. Lett. B* **2006**, *638*, 389. [CrossRef]
74. Davis, S.C.; Postma, M. Successfully combining SUGRA hybrid inflation and moduli stabilisation. *J. Cosmol. Astrop. Phys.* **2008**, *0804*, 022.
75. Shih, D. Pseudomoduli Dark Matter. *J. High Energy Phys.* **2009**, *0909*, 046. [CrossRef]
76. Dimopoulos, K.; Axenides, M. Hybrid Inflation without Flat Directions and without Primordial Black Holes. *J. Cosmol. Astrop. Phys.* **2005**, *0506*, 008. [CrossRef]
77. Planck Collaboration. Planck 2018 results. IX. Constraints on primordial non-Gaussianity. *Astron. Astrophys.* **2020**, *641*, A9. [CrossRef]
78. Maldacena, J. Non-Gaussian features of primordial fluctuations in single field inflationary models. *J. High Energy Phys.* **2003**, *0305*, 013. [CrossRef]
79. Kinney, W.H.; Vagnozzi, S.; Visinelli, L. The zoo plot meets the swampland: Mutual (in)consistency of single-field inflation, string conjectures, and cosmological data. *Class. Quant. Grav.* **2019**, *36*, 117001. [CrossRef]
80. Vafa, C. The String Landscape and the Swampland. *arXiv* **2005**, arXiv:hep-th/0509212.
81. Susskind, L. Dynamics of spontaneous symmetry breaking in the Weinberg-Salam theory. *Phys. Rev. D* **1979**, *20*, 2619. [CrossRef]
82. Bousso, R. The Cosmological Constant Problem, Dark Energy, and the Landscape of String Theory. In *Proceedings of the Subnuclear Physics: Past, Present and Future, Vatican City, Vatican, 30 October–2 November 2011*; Pontifical Academy of Sciences: Vatican City, Vatican, 2011.
83. Kumar, J. A Review of Distributions on the String Landscape. *Int. J. Mod. Phys. A* **2006**, *21*, 3441. [CrossRef]
84. Banks, T.; Fischler, W.; Shenker, S.H.; Susskind, L. M Theory As A Matrix Model: A Conjecture. *Phys. Rev. D* **1997**, *55*, 5112. [CrossRef]
85. Dijkgraaf, R.; Verlinde, E.; Verlinde, H. Matrix String Theory. *Nucl. Phys. B* **1997**, *500*, 43. [CrossRef]
86. Banks, T.; Johnson, M.; Shomer, A. A note on Gauge Theories Coupled to Gravity. *J. High Energy Phys.* **2006**, *0609*, 049. [CrossRef]
87. Randall, L.; Sundrum, R. A Large Mass Hierarchy from a Small Extra Dimension. *Phys. Rev. Lett.* **1999**, *83*, 3370. [CrossRef]
88. Randall, L.; Sundrum, R. An Alternative to Compactification. *Phys. Rev. Lett.* **1999**, *83*, 4690. [CrossRef]
89. Steinhardt, P.J. General considerations of the cosmological constant and the stabilization of moduli in the brane-world picture. *Phys. Lett. B* **1999**, *462*, 41. [CrossRef]
90. Deffayet, C.; Dvali, G.; Gabadadze, G. Accelerated Universe from Gravity Leaking to Extra Dimensions. *Phys. Rev. D* **2002**, *65*, 044023. [CrossRef]
91. Franceschini, R.; Giudice, G.F.; Giardino, P.P.; Lodone, P.; Strumia, A. LHC bounds on large extra dimensions. *J. High Energy Phys.* **2011**, *1105*, 092. [CrossRef]
92. Alcaniz, J.S.; Zhu, Z-H. Complementary Constraints on Brane Cosmology. *Phys. Rev. D* **2005**, *71*, 083513. [CrossRef]
93. Lazkoz, R.; Maartens, R.; Majerotto, E. Observational constraints on phantom-like braneworld cosmologies. *Phys. Rev. D* **2006**, *74*, 083510. [CrossRef]
94. Dubovsky, S.L.; Rubakov, V.A.; Tinyakov, P.G. Brane world: Disappearing massive matter. *Phys. Rev. D* **2000**, *62*, 105011. [CrossRef]
95. Dubovsky, S.L.; Rubakov, V.A.; Tinyakov, P.G. Is the electric charge conserved in brane world? *J. High Energy Phys.* **2000**, *8*, 041. [CrossRef]
96. Dubovsky, S.L.; Rubakov, V.A. On models of gauge field localization on a brane. *Int. J. Mod. Phys. A* **2001**, *16*, 4331. [CrossRef]

97. Dvali, G.; Gabadadze, G.; Shifman, M. (Quasi)Localized Gauge Field on a Brane: Dissipating Cosmic Radiation to Extra Dimensions? *Phys. Lett. B* **2001**, *497*, 271. [CrossRef]
98. Ziaeepour, H. Color Glass Condensate in Brane Models or Don't Ultra High Energy Cosmic Rays Probe 1015eV Scale ?. *Mod. Phys. Lett. A* **2005**, *20*, 419. [CrossRef]
99. Ziaeepour, H. QCD Color Glass Condensate Model in Warped Brane Models. *Grav. Cosmol. Suppl.* **2005**, *11*, 189.
100. Hong, D.-K.; Hsu, S.D.H. Holography, Entropy and Extra Dimensions. *Phys. Lett. B* **2004**, *591*, 208. [CrossRef]
101. Creek, S.; Gregory, R.; Kanti, P.; Mistry, B. Braneworld stars and black holes. *Class. Quant. Grav.* **2006**, *23*, 6633. [CrossRef]
102. CMS Collaboration. Search for microscopic black hole signatures at the Large Hadron Collider. *Phys. Lett. B* **2011**, *697*, 434. [CrossRef]
103. Dvali, G.; Gomez, C.; Mukhanov, S. Probing Quantum Geometry at LHC. *J. High Energy Phys.* **2011**, *1102*, 12. [CrossRef]
104. Dvali, G.; Folkerts, S.; Germani, C. Physics of Trans-Planckian Gravity. *Phys. Rev. D* **2011**, *84*, 024039. [CrossRef]
105. Dvali, G.; Giudice, G.F.; Gomez, C.; Kehagias, A. UV-Completion by Classicalization. *J. High Energy Phys.* **2011**, *08*, 108. [CrossRef]
106. Padmanabhan, T. Gravity and the Thermodynamics of Horizons. *Phys. Rep.* **2005**, *406*, 49. [CrossRef]
107. Verlinde, E.P. On the Origin of Gravity and the Laws of Newton. *J. High Energy Phys.* **2011**, *1104*, 029. [CrossRef]
108. Hu, B.L. Can Spacetime be a Condensate? *Int. J. Theor. Phys.* **2005**, *44*, 1785. [CrossRef]
109. Sindoni, L. Emergent gravitational dynamics from multi-BEC hydrodynamics?. *Phys. Rev. D* **2011**, *83*, 024022. [CrossRef]
110. 't Hooft, G. Dimensional Reduction in Quantum Gravity. *arXiv* **1993**, arXiv:gr-qc/9310026.
111. Van Raamsdonk, M. Building up spacetime with quantum entanglement. *Gen. Rel. Grav* **2010**, *42*, 2323–2329. [CrossRef]
112. Jacobson, T. Gravitation and vacuum entanglement entropy. *Int. J. Mod. Phys. D* **2012**, *21*, 1242006. [CrossRef]
113. Piazza, F. Glimmers of a pre-geometric perspective. *Found. Phys.* **2010**, *40*, 239. [CrossRef]
114. Eichhorn, A.; Koslowski, T.; Lumma, J.; Pereira, A.D. Towards background independent quantum gravity with tensor models. *Class. Quant. Grav.* **2019**, *36*, 15. [CrossRef]
115. Maldacena, J.M. The Large N Limit of Superconformal Field Theories and Supergravity. *Adv. Theor. Math. Phys* **1998**, *2*, 231. [CrossRef]
116. Donnelly, W.; Giddings, S.B. Diffeomorphism-invariant observables and their nonlocal algebra. *Phys. Rev. D* **2016**, *93*, 024030. [CrossRef]
117. Giddings, S.B. Quantum-first gravity. *Found. Phys.* **2019**, *49*, 177. [CrossRef]
118. Dirac, P.A.M. *The Principles of Quantum Mechanics*; Oxford University Press: Oxford, UK, 1958.
119. Von Neumann, J. *Mathematical Foundation of Quantum Theory*; Princeton University Press: Princeton, NJ, USA, 1955.
120. Kiefer, C. On the interpretation of quantum theory—from Copenhagen to the present day. In *Time, Quantum and Information*; Springer: Berlin/Heidelberg, Germany, 2003.

universe

MDPI

Review

Strongly Continuous Representations in the Hilbert Space: A Far-Reaching Concept

Julio Marny Hoff da Silva * and Gabriel Marcondes Caires da Rocha

Department of Physics, UNESP, Estadual Paulista University, 12516-410 Brazilia, Brazil; gabriel.marcondes@unesp.br

* Correspondence: julio.hoff@unesp.br

Abstract: We revisit the fundamental notion of continuity in representation theory, with special attention to the study of quantum physics. After studying the main theorem in the context of representation theory, we draw attention to the significant aspect of continuity in the analytic foundations of Wigner's work. We conclude the paper by reviewing the connection between continuity, the possibility of defining certain local groups, and their relation to projective representations.

Keywords: ray representation; strongly continuous; continuity; Hilbert space

Citation: da Silva, J.M.H.; da Rocha, G.M.C. Strongly Continuous Representations in the Hilbert Space: A Far-Reaching Concept. *Universe* **2021**, 7, 285. https://doi.org/10.3390/universe7080285

Academic Editors: Gerald B. Cleaver and J. W. (Jan-Willem) Van Holten

Received: 28 June 2021
Accepted: 31 July 2021
Published: 4 August 2021

1. Introduction

Since the bygone days of the 1930s, Wigner has collected strong mathematical results on mappings of continuous groups on a given Hilbert space [1]. From this time comes the famous theorem that ensures that any symmetry operation (any operation preserving probabilities) can be represented by either a linear and unitary or anti-linear and anti-unitary operator (see Appendix A). The pursuit of Poincarè group representations on Hilbert spaces, a program supported by Weyl [2], culminates in the profound and robust work celebrated here [3].

It is a complicated, if not impossible, task to pick out the most effective results from the wealth of excellent ones shown in [3]. Wigner introduced in this work the very precise notion of particle, to mention only one of the main achievements. However, the mathematical partial results collected on the way to such relevant insights are also important. We would like to revisit one of these cornerstone results in the existence of so-called admissible representatives, presenting its meaning based on continuous projective representations. Let us introduce some standard notions.

The first idea to remember is the quantum mechanical ray representation of states, which arises from the freedom that a complex field gives to states in the Hilbert space inner product. The transition probability for a quantum state ψ to turn into ϕ is expressed by $P = |(\phi,\psi)|^2$. This probability remains the same if we replace the states by rays, $\mathbf{\Psi}$. Each ray comprises an equivalence class of states containing all elements $\{\alpha\psi\}$ (where α is a unimodular complex phase), of which a given element $\alpha\psi \in \mathbf{\Psi}$, where ψ is fixed, is a representative. The inner product between two rays is defined as $\mathbf{\Phi}\cdot\mathbf{\Psi} = |(\phi,\psi)|$, where $\phi \in \mathbf{\Phi}$ and $\psi \in \mathbf{\Psi}$. In a manner akin to vector rays, Bargmann introduced the concept of operator rays [4] $\mathbf{U}$ as the set comprising all elements αU with fixed U and a unimodular complex α.

In the case where the symmetries of a physical system are described by a given continuous group H, there exists an isomorphism between each element $h \in H$ and an operator ray $\mathbf{U}_h$ such that for $h_1, h_2 \in H$, the usual representation relation (although related to operator rays) holds, i.e., $\mathbf{U}_{h_1}\mathbf{U}_{h_2} = \mathbf{U}_{h_{12}}$. These are the projective representations whose continuity will be studied here. In physics, of course, continuous representations are of great importance. In a prosaic context, for example, the regular textbooks of Quantum Mechanics and Quantum Field Theory introduce the possibility of acting on quantum states (e.g., with differential operators) without regard to possible mathematical subtleties. This and many others aspects are automatically taken as well-posed mathematically in a standard exposition of quantum physics. Thanks to the work of Wigner, in particular the aforementioned theorem, the approach presented in these textbooks is not wrong.

To justify our choice from a different framework, as as Wigner himself recognized [3], part of the generality and novelty of his approach came from mathematical rigor, in particular concerning continuity. Majorana [5] and Dirac [6] adopted continuity in several aspects of their analysis (an approach which is certainly justifiable and entirely correct, after all) but whose dubiety was completely removed by Wigner.

Now, back to the (projective) representations frame: from a mathematical point of view, phases in representations of continuous groups have become quite a sophisticated tool with the aid of algebraic topology [7,8] since the early presentation due to Bargmann. More or less recently, the original line of research gained additional interest by generalizations of the phase exponents (which enter the projective representations) involving the time (or the spacetime) parameter(s) [9,10]. In this generalization of Bargmann's theory, the selection of admissible representatives is also an important issue.

This review is devoted to an appreciation of conditions related to the existing argument of regular representatives, whose steps are also revisited. We begin in Section 2 with a recapitulation of the theorem that ensures the selection of a strongly continuous set of ray operators representatives in the Hilbert space, and discuss its standard proof (Wigner–Bargmann) step by step. Important enough, the existence of such representatives is at the very heart of our understanding of the quantum process, from spacetime evolution to probability transitions. In Section 3, we call special attention to some relevant consequences of continuity. Of course, we would not be so far saying, somewhat frivolously, that everything comes from continuity. There are, nevertheless, profound concepts linked to continuity that can be brought up in the analysis and whose full appreciation is usually bypassed. In Section 3.1, we define and investigate the continuity of local factors and study its implications to representation theory. Besides, in this section, we also place a theorem related to one-parametric group representation quite relevant in the Wigner approach. In Section 3.2, we explore how continuity also helps in the understanding of whether a given representation is projective or genuine. In the final section, we conclude. To guarantee a sequential reading of the paper we leave for the Appendix a step by step proof of Wigner's famous theorem about symmetric representations in the Hilbert space, as well as another result, relevant for the action upon connected (sub)groups.

This work is a modest contribution to the subject. We do not present new results, but give a technical appreciation to the foundations of representation theory, explicitly presenting all the proof steps and highlighting the physical interest of the results when the situation appears. Whenever possible, we follow Bargmann's exposition [4] by its didactic and far-reaching results connecting continuity to physical aspects. Throughout the paper, operators will be taken as unitary (see Proposition A1 in the Appendix A) and vector states are normalized to unity.

2. Selection of Continuous Representatives

We begin by defining standard tools for representations of continuous groups. Let $\mathcal{H}$ be a complex Hilbert space. Let the distance, $d(\mathbf{\Psi},\mathbf{\Psi}'')$, between two rays $\mathbf{\Psi}$ and $\mathbf{\Psi}''$ belonging to $\mathcal{H}$ be given by the minimum value of $||\psi-\psi''||$, where double bar stands for the usual vector norm $||\psi|| = (\psi,\psi)^{1/2}$, and ψ and ψ'' are representatives of $\mathbf{\Psi}$ and $\mathbf{\Psi}''$, respectively. This qualitative definition can be fully specified by noting that $||\psi-\psi''||^2 = 2(1-Re(\psi,\psi''))$, from which it may be proved that

$$d(\mathbf{\Psi},\mathbf{\Psi}'') = [2(1-\mathbf{\Psi}\cdot\mathbf{\Psi}'')]^{\frac{1}{2}}, \tag{1}$$

where the definition of the inner product between two rays has already been defined in the introduction. To see that Equation (1) holds, take $\mathbf{\Psi} = \sigma\psi$ and $\mathbf{\Psi}'' = \tau\psi''$, with

$|\sigma| = 1 = |\tau|$. It is always possible to find a unimodular complex number, say λ, such that $(\psi, \psi'') = \lambda|(\psi, \psi'')|$. Therefore, it is clear that

$$||\psi - \psi''||^2 = 2(1 - Re[\sigma^* \tau(\psi, \psi'')]), \quad (2)$$

where σ^* stands for the complex conjugation of σ. Note that

$$|(\psi, \psi'')| = |\sigma||\tau||(\psi, \psi'')| = |\sigma^*||\tau||(\psi, \psi'')| = |\sigma^* \tau(\psi, \psi'')| = |(\sigma\psi, \tau\psi'')| = \mathbf{\Psi} \cdot \mathbf{\Psi}'' \quad (3)$$

and therefore we can set $(\psi, \psi'') = \lambda \mathbf{\Psi} \cdot \mathbf{\Psi}''$. Returning to (2), we are left with ($\mathbf{\Psi} \cdot \mathbf{\Psi}'' \in \mathbb{R}$)

$$||\psi - \psi''||^2 = 2(1 - Re[\sigma^* \tau \lambda] \mathbf{\Psi} \cdot \mathbf{\Psi}''). \quad (4)$$

It is clear that $||\psi - \psi''||^2$ reaches its minimum for $Re[\sigma^* \tau \lambda]_{\max} = 1$, from which Equation (1) follows. As we will see in a moment, the definition of distance is crucial for the very conception of (strong) continuity for ray representations.

Definition 1. *A given ray representation of a (Lie) group H is said to be continuous if for any element $h \in H$, any $\mathbf{\Psi} \in \mathcal{H}$, and any $\epsilon > 0$, there exists a neighborhood $\mathfrak{N} \subset H$ of h such that $d(\mathbf{U}_s \mathbf{\Psi}, \mathbf{U}_h \mathbf{\Psi}) < \epsilon$, if $s \in \mathfrak{N}$.*

As an aside remark, we note that it suffices to consider the above definition for the identity element. Moreover, in a complete (metric) space, A is said to be continuous with respect to B if the set of nonzero values of A is bounded by B. In this sense, we can say that the inner product in both terms is continuous with respect to the distance $d(\mathbf{\Psi}, \mathbf{\Psi}'')$. Here is the proof: suppose four rays $\mathbf{\Psi}_1$, $\mathbf{\Psi}_2$, $\mathbf{\Phi}_1$, and $\mathbf{\Phi}_2$ belong to $\mathcal{H}$. Then,

$$\begin{aligned} |\mathbf{\Psi}_1 \cdot \mathbf{\Phi}_1 - \mathbf{\Psi}_2 \cdot \mathbf{\Phi}_2| &= |\mathbf{\Psi}_1 \cdot \mathbf{\Phi}_1 - \mathbf{\Psi}_1 \cdot \mathbf{\Phi}_2 + \mathbf{\Psi}_1 \cdot \mathbf{\Phi}_2 - \mathbf{\Psi}_2 \cdot \mathbf{\Phi}_2| \\ &\leq |(\psi_1, \phi_1) - (\psi_1, \phi_2)| + |(\psi_1, \phi_2) - (\psi_2, \phi_2)| \quad (5) \\ &= ||\phi_1 - \phi_2|| + ||\psi_1 - \psi_2||, \quad (6) \end{aligned}$$

and therefore (as the inequality always holds) we have

$$|\mathbf{\Psi}_1 \cdot \mathbf{\Phi}_1 - \mathbf{\Psi}_2 \cdot \mathbf{\Phi}_2| \leq d(\mathbf{\Phi}_1, \mathbf{\Phi}_2) + d(\mathbf{\Psi}_1, \mathbf{\Psi}_2). \quad (7)$$

Now, we will state and discuss in detail the theorem asserting the selection possibility of admissible representatives.

Theorem 1. *Let $\mathbf{U}_r$ be a continuous ray representation of a group G. For all r in a suitably chosen neighborhood $\mathfrak{N}_0$ of the unit element e of G, one can select a strongly continuous set of representatives $U_r \in \mathbf{U}_r$ (i.e., for any vector ψ, any $r \in \mathfrak{N}_0$ and any positive ϵ, there exists a neighborhood $\mathfrak{N}$ of r such that $||U_s\psi - U_r\psi|| < \epsilon$ if $s \in \mathfrak{N}$).*

The set of representatives $\{U_r\}$ satisfying these conditions is called an admissible set of representatives. From the concept of an admissible set of representatives derives a wealth of important results in representation theory in physics (ultimately including the particle concept itself). This set is indeed strongly continuous and probability transitions taken from its representatives vary continuously with the group element s, in complete agreement with the previous definition. We will discuss in detail the proof of Bargmann, who in turn followed the Wigner steps.

Proof. Let $\mathbf{\Psi}$ be a fixed ray in $\mathcal{H}$ and ψ a given representative. Define $g_r = \mathbf{\Psi} \cdot \mathbf{U}_r \mathbf{\Psi}$ for $r \in H$, and as the inner product is continuous, g_r is a continuous function of r. Therefore, it is possible to choose a suitable neighborhood $\mathfrak{N} \supset r$ such that $\alpha < g_r \leq 1$ with $\alpha \in (0, 1)$. Moreover, as a strategy for the proof, a certain representative $U_r \in \mathbf{U}_r$ is chosen such that

$$g_r = \mathbf{\Psi} \cdot \mathbf{U}_r \mathbf{\Psi} = (\psi, U_r \psi). \tag{8}$$

Note the absence of modulus[1] in (8), which is contrary to the definition of the inner product of vector rays. We will address this point after completing the standard proof Until then, we will only emphasize that $e \in \mathfrak{N}$, just as it is contained in the statement.

Let $\psi \in \mathbf{\Psi}$, $r, s \in \mathfrak{N}$ and define the quantities (partially preserving the Bargmann notation)

$$d_{r,s}(\psi) = d(\mathbf{U}_r \mathbf{\Psi}, \mathbf{U}_s \mathbf{\Psi}), \tag{9}$$

$$\sigma_{r,s}(\psi) = (U_r \psi, U_s \psi), \tag{10}$$

$$Z_{r,s}(\psi) = U_s \psi - \sigma_{r,s}(\psi) U_r \psi. \tag{11}$$

These quantities will help the proof process. Note that $Z_{r,s}(\psi)$ is orthogonal to $U_r \psi$, as can be easily seen from

$$(U_r \psi, Z_{r,s}(\psi)) = (U_r \psi, U_s \psi) - \sigma_{r,s}(\psi), \tag{12}$$

which vanishes by means of (10). From this, one can see that

$$||Z_{r,s}(\psi)||^2 = (U_s \psi - \sigma_{r,s}(\psi) U_r \psi, U_s \psi - \sigma_{r,s}(\psi) U_r \psi) = 1 - \sigma_{r,s}(\psi)(U_s \psi, U_r \psi) \tag{13}$$

and therefore (again using (10))

$$||Z_{r,s}(\psi)||^2 = 1 - |\sigma_{r,s}(\psi)|^2. \tag{14}$$

It follows straightforwardly from Equation (1) that $1 - |(\psi, \phi)|^2 \leq d^2$, leading to

$$||Z_{r,s}(\psi)||^2 \leq d^2_{r,s}(\psi). \tag{15}$$

Now, taking $\psi = \phi$ and calculating $(\phi, Z_{r,s}(\phi))$, we have

$$\sigma_{r,s}(\phi) = \frac{1}{g_r}[g_s - (\phi, Z_{r,s}(\phi))]. \tag{16}$$

Recalling that $||\psi - \psi''||^2 = 2(1 - Re(\psi, \psi''))$ we have[2] $||\psi - \psi''||^2 \leq 2|1 - (\psi, \psi'')|$. Therefore,

$$||U_s \phi - U_r \phi||^2 \leq 2|1 - (U_r \phi, U_s \phi)| = 2|1 - \sigma_{r,s}(\phi)| \tag{17}$$

and by (16) we find

$$||U_s \phi - U_r \phi||^2 \leq 2\left|\frac{1}{g_r}[g_r - g_s + (\phi, Z_{r,s}(\phi))]\right|. \tag{18}$$

As $g_r > \alpha$, it is possible to rewrite the above equation as

$$||U_s \phi - U_r \phi||^2 \leq \frac{2}{\alpha}|g_r - g_s + (\phi, Z_{r,s}(\phi))| \leq \frac{2}{\alpha}[|g_r - g_s| + |(\phi, Z_{r,s}(\phi))|]. \tag{19}$$

As defined before, the functions g_r give $|g_r - g_s| = |(\phi, U_r \phi) - (\phi, U_s \phi)| = |(\phi, U_r \phi - U_s \phi)| \leq ||U_r \phi - U_s \phi||$. Therefore, $|g_r - g_s|$ is less than (or equal to) any value of $||U_r \phi - U_s \phi||$; in particular the inequality holds for the minimum value of $||U_r \phi - U_s \phi||_{\min} = d_{r,s}(\phi)$. Therefore $|g_r - g_s| \leq d_{r,s}(\phi)$. Moreover, $|(\phi, Z_{r,s}(\phi))| \leq ||Z_{r,s}(\phi)||$ and using (15) we have $|(\phi, Z_{r,s}(\phi))| \leq d_{r,s}(\phi)$. Collecting all these results, we finally get

$$||U_s \phi - U_r \phi||^2 \leq \frac{4}{\alpha} d_{r,s}(\phi), \tag{20}$$

which ensures continuity for $U_r\phi$ in the sense of the highlighted definition before the theorem. The next step is to ensure continuity for a vector χ given by $\chi = (\phi + \varphi)/\sqrt{2}$, where the normalized vector φ is assumed to be orthogonal[3] to ϕ. Note that

$$(U_r\phi, Z_{r,s}(\chi)) = (U_r\phi, U_s\chi) - \sigma_{r,s}(\chi)(U_r\phi, U_r\chi) \tag{21}$$

now adding and subtracting (U_sh, U_sk) to (21), the result may be recast as

$$(U_r\phi, Z_{r,s}(\chi)) = (U_r\phi - U_s\phi, U_s\chi) + (U_s\phi, U_s\chi) - \sigma_{r,s}(\chi)(U_r\phi, U_r\chi). \tag{22}$$

In turn, $(U_m\phi, U_m\chi) = (\phi, U_m^\dagger U_m\chi) = (\phi, \frac{1}{\sqrt{2}}[\phi + \varphi]) = 1/\sqrt{2}$ for every $m \in H$, in particular for $m \in \mathfrak{N} \subset H$. Returning to (21), then we have

$$(U_r\phi, Z_{r,s}(\chi)) + (U_s\phi - U_r\phi, U_s\chi) = \frac{1}{\sqrt{2}}(1 - \sigma_{r,s}(\chi)). \tag{23}$$

Now we can apply Equations (23) to (17) (suitable adequate to χ) and arrive at

$$||U_s\chi - U_r\chi||^2 \leq 2^{2/3}\Big\{|(U_r\phi, Z_{r,s}(\chi))| + |(U_s\phi - U_r\phi, U_s\chi)|\Big\}. \tag{24}$$

Both terms of the right-hand side of Equation (24) are bounded from above by Schwarz inequality. Then, using Equation (15) in the first term, we have

$$||U_s\chi - U_r\chi||^2 \leq 2^{2/3}\Big\{d_{r,s}(\chi) + ||U_s\phi - U_r\phi||\Big\} \tag{25}$$

and the continuity of $U_r\phi$ implies the continuity of $U_r\chi$ (and of course of $U_r\varphi$). Finally, if ψ is a linear combination written in terms of ϕ and φ, as defined earlier, then $U_r\psi$ is clearly continuous. □

As a final remark before going further, note that the first steps of the proof could be repeated around any group element $k \in H$ by simply adapting the neighborhood to include k and starting with the definition of g_r functions as $g_r = \mathbf{U}_k\mathbf{\Psi} \cdot \mathbf{U}_r(\mathbf{U}_k\mathbf{\Psi})$.

Additional Discussion

Apart from the comments on the proof inserted here and there, let us concentrate on the determination of the g_r functions. First, a fact: there is no loss of generality in choosing a neighborhood of $e \in H$ such that $g_r > \alpha$ for $\alpha \in (0,1)$ or even in setting the operator such that $g_r = (\psi, U_r\psi)$ instead of $|(\psi, U_r\psi)|$. The continuity of the inner product with respect to the distance ensures this last procedure. However, it is instructive, to see the effect of such a device within the proof scheme when the necessity emerges, as it were, rather than *a priori*.

Let us begin with functions $\tilde{g}_r = \mathbf{\Psi} \cdot \mathbf{U}_r\mathbf{\Psi} = |(\psi, U_r\psi)|$ for which the group separability condition $\tilde{g}_r > \alpha$ for $r \in \mathfrak{N} \subset H$ holds. Given this definition, part of the procedure used in the previous proof is unhelpful. We first note that with $Z_{r,s}(\phi)$, as defined in (11),

$$|(\phi, Z_{r,s}(\phi))| = |(\phi, U_s\phi) - \sigma_{r,s}(\phi)(\phi, U_r\phi)| \geq |(\phi, U_s\phi)| - |\sigma_{r,s}(\phi)||(\phi, U_r\phi)|, \tag{26}$$

using the standard triangle inequality for complex numbers. In a more compact form

$$|(\phi, Z_{r,s}(\phi))| \geq \tilde{g}_s - |\sigma_{r,s}(\phi)|\tilde{g}_r. \tag{27}$$

From (27), we read

$$|1 - |\sigma_{r,s}(\phi)|| \leq \frac{1}{\alpha}\Big\{|\tilde{g}_r - \tilde{g}_s| + |(\phi, Z_{r,s}(\phi))|\Big\}. \tag{28}$$

Again, both terms in the right-hand side of (28) are bounded from above by the distance in the Hilbert space. In the second term, Equation (15) will be used again, while for the first term we have

$$|\tilde{g}_r - \tilde{g}_s| = ||(\phi, U_r\phi)| - |(\phi, U_s\phi)|| \leq |g_r - g_s| \tag{29}$$

and the discussion around (19) holds. Therefore,

$$|1 - |\sigma_{r,s}(\phi)|| \leq \frac{2}{\alpha} d_{r,s}(\phi). \tag{30}$$

Now recall that a direct computation leads to Equation (17), i.e.,

$$||U_s\phi - U_r\phi||^2 \leq 2|1 - \sigma_{r,s}(\phi)| \tag{31}$$

and thus the procedure to ensure continuity of $U_r\phi$ requires additional attention, since $|1 - \sigma_{r,s}(\phi)| \geq |1 - |\sigma_{r,s}(\phi)||$. Let us examine both cases.

In the case that $|1 - \sigma_{r,s}(\phi)| > |1 - |\sigma_{r,s}(\phi)||$ one is not able to compare $||U_s\phi - U_r\phi||^2$ and $d_{r,s}(\phi)$ accurately. This situation then leads to an empty tautology: the selection of continuous representatives is the one that choose $U_r\phi$ respecting $||U_s\phi - U_r\phi||^2 \leq (4/\alpha)d_{r,s}(\phi)$, i.e., the continuous one.

However, when $|1 - \sigma_{r,s}(\phi)| = |1 - |\sigma_{r,s}(\phi)||$, one can indeed claim continuity for $U_r\phi$ as can be easily seen. This condition (leading to the proof) is mathematically satisfied whenever $Re(\sigma_{r,s}(\phi)) \geq 0$ and $Im(\sigma_{r,s}(\phi))$ vanishes. However, these conditions are precisely the conditions studied when setting g_r functions (without tilde). Indeed,

$$\sigma_{r,s}(\phi) = (U_r\phi, U_s\phi) = (\phi, U_r^\dagger U_s\phi) = (\phi, U_{r^{-1}s}\phi), \tag{32}$$

is nothing but a g_m function for $m = r^{-1}s \in \mathfrak{N}$. Therefore, given the conditions $Re(g_m(\phi)) \geq 0$ and $Im(g_m(\phi)) = 0$ we have $g_m = \tilde{g}_m$ leading to[4] $||U_s\phi - U_r\phi||^2 \leq (4/\alpha)d_{r,s}(\phi)$. Going further, by calling $U_m = \tau U_m^0$, $\phi = \delta\phi^0$ for fixed U_m^0 and ϕ^0 and denoting $(\phi^0, U_m^0\phi^0) = X + iY$ and $\tau = \tau_1 + i\tau_2$ $(X, Y, \tau_1, \tau_2 \in \mathbb{R})$, the conditions give.

$$\begin{aligned} X\tau_1 - Y\tau_2 &\geq 0, \\ X\tau_2 + Y\tau_1 &= 0. \end{aligned} \tag{33}$$

It is not difficult to satisfy Equation (33) without any operator particularization. The last additional remark is that in the absence of a phase in the representation, i.e., for genuine (not projective) representations, continuity for the representatives is ultimately attainable by continuity of the inner product.

3. Consequences of Continuity

We will here point out some direct consequences of continuity, relevant to the mathematical structure underlining the understanding of quantum physics, first examining local factors and then investigating some consequences of the selection of continuous representatives together with freedom in the selection procedure.

3.1. Continuous Local Factors

Let us assume implicitly in what follows that all group elements to be worked out in this section belong to the same neighborhood $\mathfrak{N}$ (or to suitable intersections of neighborhoods), so that the group operations are locally well-defined. This requires $e \in H$ to be an element of the neighborhood. An admissible set of representatives engenders a continuous ray representation in $\mathfrak{N}$ and since U_rU_s and U_{rs} belongs to the same ray, we have

$$U_rU_s = \omega(r, s)U_{rs}, \tag{34}$$

where $|\omega(r,s)| = 1$ and clearly $\omega(r,e) = \omega(e,s) = \omega(e,e) = 1$, as $U_e = 1$. Note that the associative law of the group representation implies

$$\omega(r,s)\omega(rs,m) = \omega(s,m)\omega(r,sm). \tag{35}$$

The functions $\omega(r,s)$ are the so-called local factors of a given ray representation and the continuity of admissible representatives leads to the continuity of them. Let us demonstrate this fact.

Lemma 1. *For an admissible set of representatives, the local factors are continuous.*

Proof. We start with a simple truism. Let $\psi \in \mathcal{H}$, then obviously

$$[\omega(r',s') - \omega(r,s)]U_{r's'}\psi = \omega(r',s')U_{r's'}\psi - \omega(r,s)U_{r's'}\psi. \tag{36}$$

Now, adding and subtracting the terms $U_{r'}U_s\psi$ and $U_rU_s\psi$ we obtain, using (34),

$$[\omega(r',s') - \omega(r,s)]U_{r's'}\psi = \omega(r,s)(U_{rs} - U_{r's'})\psi + U_{r'}(U_{s'} - U_s)\psi + (U_{r'} - U_r)U_s\psi. \tag{37}$$

Thus, we arrive at

$$\begin{aligned}||[\omega(r',s') - \omega(r,s)]U_{r's'}\psi|| \leq ||\omega(r,s)(U_{rs} - U_{r's'})\psi||+ \\ + ||U_{r'}(U_{s'} - U_s)\psi|| + ||(U_{r'} - U_r)U_s\psi||.\end{aligned} \tag{38}$$

The left-hand side of (38) simplifies[5] to $|\omega(r',s') - \omega(r,s)|$. Moreover, since $|\omega(r,s)| = 1$ and U_r is unitary, the terms of the right-hand side are easy to handle. Thus, calling $U_{s'}\psi = \psi'$, we have

$$|\omega(r',s') - \omega(r,s)| \leq ||(U_{rs} - U_{r's'})\psi|| + ||(U_s - U_{s'})\psi|| + ||(U_{r'} - U_r)\psi'||, \tag{39}$$

from which the local factors for admissible representatives are indeed continuous. □

Continuity of phase factors can also be used to treat differentiability precisely [4]. We will just note here that a multidimensional Lie group H has elements in bijective correspondence with open balls containing $\mathfrak{N}$ in an Euclidean space of same dimensionality [11]. To fix ideas, let us consider a fixed element s and think of $r = r(r^1, r^2, \ldots, r^n)$, where $n = \dim(H)$ and r^i $(i = 1, \ldots, n)$, as the coordinates of r. The unity for the local factor may be written as $1 = \omega(e,s)$ and the unity element $e \in H$ has coordinates given by $(0,0,\ldots,0)$. It follows that

$$|\omega(r,s) - 1| = |\omega(0 + r^1, 0 + r^2, \ldots, 0 + r^n, s) - \omega(0,0,\ldots,0,s)| : \mathfrak{N} \subset \mathbb{R}^n \to \mathbb{R}^+. \tag{40}$$

For the argument, if we take $s' = e = r'$ in (39) and denote $||(1 - U_m)\psi|| \leq \kappa d_{e,m}^{1/2}$ where κ stands for a constant (a form certainly valid for admissible representatives), then (39) reads

$$|\omega(0 + r^1, 0 + r^2, \ldots, 0 + r^n, s) - \omega(0,0,\ldots,0,s)| \leq \kappa \sum_{i=r,s,rs} d_{e,i}^{1/2} \tag{41}$$

and the continuity of admissible representatives naturally bounds the local factor from above, in the sense just described.

So far we have said that the local factors are group elements dependent. It is easy to see that there is no dependence of local factors with respect to the state on which the group elements are represented [12]. Let ψ_i $(i = 1,2)$ be two linearly independent vectors in $\mathcal{H}$ and assume a dependence of the local phases on the states; also assume $\psi_3 = \psi_1 + \psi_2$. It follows that $U_r U_s \psi_3 = \omega_3(r,s) U_{rs}\psi_3$. However (recall that the operators acting in $\mathcal{H}$ are unitary and linear),

$$\omega_3(r,s)U_{rs}(\psi_1 + \psi_2) = \omega_1(r,s)U_{rs}\psi_1 + \omega_2(r,s)U_{rs}\psi_2 \tag{42}$$

and acting from the left in (42) with $U_{s^{-1}r^{-1}}$ we have $\omega_3(r,s) = \omega_1(r,s) = \omega_2(r,s)$, so no state dependence at all.

Before further studying the local factors and their continuity, we make a parenthetical but important remark about continuity in irreductible representations and one-parametric Lie subgroups. A local one-parameter Lie subgroup is a continuous curve $g = g(\lambda)$ in $\mathfrak{N}$ (λ takes values in an open real interval) with $g(\lambda_1)g(\lambda_2) = g(\lambda_1 + \lambda_2)$ for $\lambda_1, \lambda_2, \lambda_1 + \lambda_2$ defined in the same open interval in $\mathbb{R}$. In view of this remark, we can state the following theorem [3].

Theorem 2. *Let $U_{g(\lambda)}$ be an infinitesimal operator of a continuous one-parametric subgroup. If there exists a physical state, say Ψ, on which the application of $U_{g(\lambda)}$ is well defined, then there exists an everywhere dense set of such states for irreducible representations.*

Proof. If there exists such a Ψ state, then the limit $\lim_{\lambda\to 0} \lambda^{-1}(U_{g(\lambda)} - 1)\Psi$ is clearly well defined. Now, let S be a given operator of the representation. The one-parametric property $U_{g(\lambda_1)}U_{g(\lambda_2)} = U_{g(\lambda_1+\lambda_2)}$ also holds for $S^{-1}U_{g(\lambda)}S$ and therefore the limit

$$\lim_{\lambda\to 0} \frac{1}{\lambda}(S^{-1}U_{g(\lambda)}S - 1)\Psi \tag{43}$$

is also well defined. Taking $1 = S^{-1}S$ in the above expression, we see that the following limit is also well defined

$$\lim_{\lambda\to 0} \frac{1}{\lambda}(U_{g(\lambda)} - 1)S\Psi. \tag{44}$$

This means that as far as the infinitesimal operator can be applied to Ψ, it can also be applied to $S\Psi$. Thus, for irreductible representations, the existence of $\Psi \in \mathcal{H}$, which implies (44), means that for every state on which infinitesimal $U_{g(\lambda)}$ can act, every $\mathfrak{V} \subset \mathcal{H}$ contains at least one $S\Psi$ that has exactly this property, giving rise to an everywhere dense set of such states in $\mathcal{H}$. □

Just as an additional remark about the theorem just proved, if ψ does not belong to an irreducible representation, then one cannot claim the existence of such an everywhere dense set (see in [3] for further discussion). Finally, this theorem states that infinitesimal operators in the Hilbert space can be treated in a somewhat ordinary way.

3.2. *Exploring the Continuity of Representatives*

The continuity of representatives and local factors proved so far can be summarized in the expressions (25) and (39). It is now important to consider this strong continuity property together with the dependence of local factors on the choice of representatives. To begin with, consider $|\phi(r)| = 1$ and select representatives in the same ray such that $U'(r) = \phi(r)U_r$, $\forall r \in \mathfrak{N} \subset H$. Of course, $U'(r)U'(s) = \omega'(r,s)U'(rs)$, which shows that the $\omega(r,s)$ function depends on the selection of representatives, given that

$$\omega'(r,s) = \omega(r,s)\frac{\phi(r)\phi(s)}{\phi(rs)}. \tag{45}$$

This freedom in the representative selection is obviously inherited from the ray representation and its systematic study is very informative as it reveals the deep relationship between the representation itself and the group being represented. Before we begin to explore this freedom however, let us specify the analysis to the case of interest. Let $\{U'_r = \phi(r)U_r\}$ be an admissible set of representatives defined in a suitable neighborhood of e. The strong continuity of U'_r and U_r naturally implies that $\phi(r)$ is a continuous complex unimodular function. On the other hand, starting from $\phi(r)$ continuous and U_r strongly continuous, one arrives at U'_r strongly continuous.[6]

Proposition 1. *Let ω be a local factor defined in a given neighborhood $\mathfrak{N}$ of e and let $\phi(r)$ be a continuous unimodular complex function such that $\phi(e) = 1$ and the function ω' is given as in (45). Then, the function ω' is also a local factor.*

Proof. As noted earlier, (1) the strong continuity of admissible representatives implies that $\phi(r)$ is a continuous function. Therefore, ω' also satisfies (39) for admissible U'_r. Moreover, (2) $\omega'(r,e) = \omega'(e,s) = 1 = \omega'(e,e)$ as can be easily seen. Finally, (3) using (45), it is possible to write

$$\omega'(r,s)\omega'(rs,m) = \omega(r,s)\omega(rs,m)\frac{\phi(r)\phi(s)\phi(m)}{\phi(rsm)} \tag{46}$$

and using (35) we have

$$\omega'(r,s)\omega'(rs,m) = \omega(s,m)\omega(r,sm)\frac{\phi(r)\phi(s)\phi(m)}{\phi(rsm)}. \tag{47}$$

Multiplying the right-hand side of (47) by $1 = \phi(sm)/\phi(sm)$ and rearranging the terms, we obtain

$$\omega'(r,s)\omega'(rs,m) = \omega(s,m)\frac{\phi(s)\phi(m)}{\phi(sm)}\omega(r,sm)\frac{\phi(r)\phi(sm)}{\phi(rsm)}, \tag{48}$$

showing that (35) then holds also for ω', leading to a local factor. □

We will henceforth study the freedom in selecting admissible representatives on a more comprehensive basis. To do so, it may be a good time to introduce a notation with which the reader is probably more familiar. Let $\delta(r,s)$ be a strongly continuous real function defined for every group element in a suitable chosen neighborhood, satisfying the condition $\delta(e,e) = 0$ and

$$\delta(r,s) + \delta(rs,m) = \delta(s,m) + \delta(r,sm). \tag{49}$$

Then, it is possible to replace the local factors by the so-called local exponents by $\omega(r,s) = e^{i\delta(r,s)}$ (note that (49) is the counterpart of the group associative law to the local exponents).

Proposition 2. *For every local exponent, the relations (1) $\delta(r,e) = 0 = \delta(e,s)$ and (2) $\delta(r,r^{-1}) = \delta(r^{-1},r)$ hold.*

Proof. For (1) just take $s = m = e$ ($r = s = e$) in (49) and recall that $\delta(e,e) = 0$ to get $\delta(r,e) = 0$ ($\delta(e,s) = 0$). For (2) take $s = r^{-1}$ in (49) and obtain

$$\delta(r,r^{-1}) + \delta(rr^{-1},m) = \delta(r^{-1},m) + \delta(r,r^{-1}m) \tag{50}$$

and then take $m = r$ to obtain $\delta(r,r^{-1}) = \delta(r^{-1},r)$. □

When studying of physical representations, it is relevant to understand when (and how) projective representations may be discarded and one can work directly with genuine representations. This is a subtle aspect, whose answer may go beyond non-trivial points of mathematical theory. We will here appreciate only some of the aspects that are somehow directly related to continuity. In any case, the following definition is crucial for a proper exposition of this topic.

Definition 2. *Let $x(r)$ be a continuous real function defined in a neighborhood $\mathfrak{K}$ that includes products of group elements, and let δ and δ' be local exponents defined in $\mathfrak{N}$ and $\mathfrak{N}'$, respectively. Let it be assumed that $\mathfrak{K} \subset (\mathfrak{N} \cap \mathfrak{N}')$. The local exponents δ and δ' will be called equivalent if the relation*

$$\delta'(r,s) = \delta(r,s) + \Delta_{r,s}[x], \tag{51}$$

with

$$\Delta_{r,s}[x] = x(r) + x(s) - x(rs) \tag{52}$$

holds in $\mathfrak{K}$.

From Definition 2 above, it is easy to see that $x(e) = 0$ insofar as $\delta'(e,e) = \delta(e,e)$. The functional form of $\Delta_{r,s}[x]$ may be, perhaps, better justified by noting from (45) that two equivalent local exponents uniquely define two equivalent local factors with $\phi(r) = e^{ix(r)}$. By slightly changing the order of exposition, it is possible to enunciate the following proposition.

Proposition 3. *If δ is a local exponent defined in a given neighborhood, and $x(r)$ a continuous real function such that $x(e) = 0$ defined in a suitable neighborhood (see Definition 2), then $\delta'(r,s)$ as defined by (51) and (52) is a local exponent.*

Proof. First, note that the continuity of δ and x guarantees continuity for δ'. Moreover, as $x(e) = 0$, then $\delta'(e) = 0$ directly. Besides

$$\delta'(r,s) + \delta'(r,sm) = \delta(r,s) + \delta(rs,m) + \Delta_{r,s}[x] + \Delta_{rs,m}[x] \tag{53}$$

and by means of (49)

$$\delta'(r,s) + \delta'(r,sm) = \delta(s,m) + \delta(r,sm) + \Delta_{r,s}[x] + \Delta_{rs,m}[x]. \tag{54}$$

Using (51) we now arrive at

$$\delta'(r,s) + \delta'(r,sm) = \delta'(s,m) + \delta'(r,sm) + \Delta_{r,s}[x] + \Delta_{rs,m}[x] - \Delta_{s,m}[x] - \Delta_{r,sm}[x]. \tag{55}$$

Finally, note that the sum of Δ's vanishes identically. □

Before proceeding to study the consequences of continuity, we will make some complementary observations. Let us denote the equivalence between two local exponents by $\delta' \sim \delta$. This equivalence relation is

(i) Symmetric: $\delta' \sim \delta$ means $\delta'(r,s) = \delta(r,s) + \Delta_{r,s}[x]$. Therefore $\delta \sim \delta'$ as $x \to -x$;
(ii) Reflexive: obviously $\delta' \sim \delta'$;
(iii) Transitive: suppose $\delta_1(r,s) = \delta(r,s) + \Delta_{r,s}[x_1]$ in $\mathfrak{N}_1$ and $\delta_2(r,s) = \delta_1(r,s) + \Delta_{r,s}[x_2]$ in $\mathfrak{N}_2$. Then, $\delta_2(r,s) = \delta(r,s) + \Delta_{r,s}[x_1 + x_2]$ in some $\mathfrak{N} \subset (\mathfrak{N}_1 \cap \mathfrak{N}_2)$, and thus $\delta_2 \sim \delta$. Therefore, (51) and (52) set a formal equivalence class, a truly equivalence class indeed.

The observation of the last paragraph can be complemented by the following remark: if δ_1 and δ_2 are local exponents in $\mathfrak{N}_1$ and $\mathfrak{N}_2$, respectively, then every linear combination $\kappa_1\delta_1(r,s) + \kappa_2\delta_2(r,s)$ with $\kappa_i \in \mathbb{R}$ $(i = 1,2)$ is also a local exponent in $\mathfrak{N} \subset (\mathfrak{N}_1 \cap \mathfrak{N}_2)$. Summarizing these observations, we can claim that the equivalence class of a linear combination depends only on the equivalence classes of local exponents entering in the linear combination.

This is a good point to appreciate an important theorem due to Weyl [2].

Theorem 3. *For a finite dimensional continuous ray representation $\mathbf{U}_r$ of a group H, every local factor is equivalent to 1.*

Proof. Starting from $U_r U_s = \omega(r,s)U_{rs} = e^{i\delta(r,s)}U_{rs}$, assuming that n is the dimension of the representation space, and taking the determinant of the above expression, we are left with

$$\det U_r \det U_s = e^{in\delta(r,s)} \det U_{rs}. \tag{56}$$

Now, as U_r is strongly continuous, it follows that $\det U_r$ is a continuous function of r. Furthermore, $|\det U_r| = 1 = \det U_e$. Thus in a suitable neighborhood, it is possible to write $\det U_r \equiv e^{i\Sigma(r)}$ with continuous real functions such that $\Sigma(e) = 0$. In this vein, we have from (56) that

$$\delta(r,s) - \frac{\Sigma(r)}{n} - \frac{\Sigma(s)}{n} + \frac{\Sigma(rs)}{n} = 0 \tag{57}$$

and as we recognize $x(r) = -\Sigma(r)/n$ we have $\delta(r,s) + \Delta_{r,s}[x = -\Delta/n] = 0 \sim \delta'(r,s)$, from which $\omega(r,s) \sim 1$ follows. □

This is a remarkable result, which makes it clear that continuity of representatives acting in Hilbert spaces is not only a pleasant and desirable property but can also constrain important aspects of the representation that would otherwise be unspecified. However, two crucial limitations of the previous result should be noted: first, it is a local achievement, which for this reason is valid in a given neighborhood $\mathfrak{N}$ of e. To extend such claim to all group manifold, we should be able to demonstrate it for $\mathfrak{N} = H$. Notably, this is only possible if the zeroth and first homotopy groups of the manifold associated to H, $\pi_0(H)$, and $\pi_1(H)$, are both trivial[7] [14]. When these requirements are not satisfied, some additional subtitles may appear, such as in the case of representations up to a sign for $SO(3) = SU(2)/\mathbb{Z}_2$ rotations, which are of great impact in spinorial representations. The second point to emphasize is that the last theorem deals with finite-dimensional representations. We will continue the analysis by lifting this restriction and further investigating continuous representations on general basis.

As a typical representation, the operators of continuous operator rays form a group under multiplication. This concept can be systematized by introducing of the so-called local group L, which entails the freedom in selecting a given operator within a ray and formalizes, so to speak, the observation with which we started this section. To introduce this group, note that an operator belonging to an admissible set of representatives defined in a suitable neighborhood is given by $e^{i\sigma}U_r$, where the representation continuity fixes the range of σ as the real numbers. Therefore,

$$(e^{i\sigma_1}U_r)(e^{i\sigma_2}U_s) = e^{i(\sigma_1+\sigma_2)}U_r U_s = e^{i(\sigma_1+\sigma_2)}\omega(r,s)U_{rs}, \tag{58}$$

or, in terms of local exponents,

$$(e^{i\sigma_1}U_r)(e^{i\sigma_2}U_s) = e^{i(\sigma_1+\sigma_2+\delta(r,s))}U_{rs}. \tag{59}$$

We call $\mathfrak{N}^2$ the neighborhood comprising the products of any two group elements belonging to $\mathfrak{N}$, and require that $\mathfrak{N}^2$ is a neighborhood of e, and define L to be the set of elements of the form $[\sigma, r]$ with $\sigma \in \mathbb{R}$ and $r \in \mathfrak{N}$.

Now define a product $\diamond : L \times L \to L$ such that

$$[\sigma_1, r] \diamond [\sigma_2, s] = [\sigma_1 + \sigma_2 + \delta(r,s), rs], \tag{60}$$

where $\delta(r,s)$ is a local exponent, in full agreement with (59) but bypassing any allusion to a given representative (provided it is admissible).

Proposition 4. *L is a group under $\diamond$.*

Proof. (1) As it can be readily seen, the unity element is simply given by $[0, e]$. (2) Note that

$$[\sigma_1, r_1] \diamond \Big([\sigma_2, r_2] \diamond [\sigma_3, r_3]\Big) = [\theta_1 + \theta_2 + \theta_3 + \delta(r_2, r_3) + \delta(r_1, r_2 r_3), r_1 r_2 r_3], \tag{61}$$

while

$$\Big([\sigma_1, r_1] \diamond [\sigma_2, r_2]\Big) \diamond [\sigma_3, r_3] = [\theta_1 + \theta_2 + \theta_3 + \delta(r_1, r_2) + \delta(r_1 r_2, r_3), r_1 r_2 r_3]. \tag{62}$$

Thus, the associative aspect of the representation inherited by local factors (49) yields an associative product. (3) For every element of L, the unity element is reachable by a composition with $[\sigma, r]^{-1} \equiv [-(\sigma + \delta(r, r^{-1})), r^{-1}]$, where $r^{-1} \in \mathfrak{N}$, for

$$\begin{aligned} [\sigma, r] \diamond [\sigma, r]^{-1} &= [\sigma, r] \diamond [-(\sigma + \delta(r, r^{-1})), r^{-1}] \qquad (63) \\ &= [\sigma - \sigma - \delta(r, r^{-1}) + \delta(r, r^{-1}), r r^{-1}] \\ &= [0, e] = [\sigma, r]^{-1} \diamond [\sigma, r]. \qquad (64) \end{aligned}$$

This concludes the proof. □

Then, the group L has $\mathbb{R} \times \mathfrak{N}^2$ as associated manifold and it is often said that L is the local group constructed for the local exponent δ. Despite the apparent simplicity of the local group, its structure is relevant enough to be analyzed further. Take elements of the form $[\sigma, e]$ of L. Of course, these elements form an one-parameter subgroup, say C, of L. Let us present some properties of C that can be easily checked. First, C belongs to the center of L. In fact $[\sigma, e] \diamond [\alpha, r] = [\sigma + \alpha + \delta(e, r), er]$, but $\delta(e, r) = 0 = \delta(r, e)$ as it is a local exponent and of course $er = re = r$. Thus, $[\sigma, e] \diamond [\alpha, r] = [\alpha, r] \diamond [\sigma, e]$, $\forall\, [\alpha, r] \in L$. Now, it is not hard to prove that $[\sigma, r] = [\sigma, e] \diamond [0, r]$, and then every element of L can be written in terms of an element belonging to C (something relevant in what follows) together with an element of H.

The central group investigation is important to understand a relevant aspect between the center of algebras and the local existence of projective representations. In fact, by inspecting elements of C one sees that it comprises all the information about the freedom in a ray selection. To see that the inspection of L (and therefore C) is indeed informative about representations of H, let us show some relevant isomorphisms.

Lemma 2. *The quotient group L/C is locally isomorphic to H.*

Proof. As known, the group L/C has elements belonging to the set $\{\kappa C | \kappa \in L\}$, while H comprises elements $r_1, r_2, \ldots$. Let φ be an application from $L/C \to H$. Before setting φ completely, we note that $\varphi(\kappa C)$ typically has $\varphi([\sigma_1, r] \diamond [\sigma_2, e])$ as arguments. It turns out that

$$[\sigma_1, r] \diamond [\sigma_2, e] = [0, r] \diamond [\sigma_1, e] \diamond [\sigma_2, e] = [0, r] \diamond [\sigma_1 + \sigma_2, e]. \tag{65}$$

Then, calling $\sigma_1 + \sigma_2 \equiv \sigma$ we have $[\sigma_1, r] \diamond [\sigma_2, e] = [0, r] \diamond [\sigma, e]$. The definition of φ is thus completed by selecting the pure H element of its domain, that is $\varphi([0, r] \diamond [\sigma, e]) \doteq r$. In view of this we see that (1) $\varphi(e) = \varphi([0, e] \diamond [0, e]) = \varphi([0, e])$ and hence $\varphi([0, e]) \doteq e = \varphi(e)$. Now, note that

$$\varphi\Big([0, r]^{-1} \diamond [\sigma, e]^{-1}\Big) = \varphi\Big([-\delta(r, r^{-1}), r^{-1}] \diamond [-(\sigma + \delta(e, e)), e]\Big), \tag{66}$$

where we have used $e^{-1} = e$. As $\delta(e, r) = 0$, $\forall r$ locally, we have

$$\begin{aligned} \varphi\Big([0, r]^{-1} \diamond [\sigma, e]^{-1}\Big) &= \varphi\Big([-\delta(r, r^{-1}), r^{-1}] \diamond [-\sigma, e]\Big) \\ &= \varphi\Big([-\sigma - \delta(r, r^{-1}), r^{-1}]\Big) \doteq r^{-1}. \end{aligned} \tag{67}$$

Finally, since $\varphi([0, r] \diamond [\sigma, e]) = r$, we arrive at $\varphi(g^{-1}) = \varphi^{-1}(g)\ \forall g \in L/C$. □

Incidentally, we might note that $\varphi([\sigma_1, e] \diamond [0, r_1] \diamond [\sigma_2, e] \diamond [0, r_2]) = \varphi([\sigma_1 + \sigma_2 + \delta(r_1, r_2), r_1 r_2]) \doteq r_1 r_2$. However, obviously, $\varphi([\sigma_i, e] \diamond [0, r_i]) \doteq r_i$ $(i = 1, 2)$ and then $\varphi(g_1 g_2) = \varphi(g_1)\varphi(g_2), \forall g_1, g_2, g_1 g_2 \in L/C$.

Besides the important result just described in Lemma 2, it is also possible to locally connect local groups with equivalent exponents. This is the content of the next lemma.

Lemma 3. *Let δ and $\tilde{\delta}$ be two equivalent local exponents in a given neighborhood, that is $\tilde{\delta}(r, s) = \delta(r, s) + \Delta_{rs}[x]$. Then, the corresponding local groups L and $\tilde{L}$ are locally isomorphic.*

Proof. Consider the mapping $\varphi : L \to \tilde{L}$ such that $\varphi([\sigma, r]) = [\sigma - x(r), \tilde{r} = r]$. It is clear that

$$\begin{aligned} \varphi([\sigma_1, r_1]) \diamond \varphi([\sigma_2, r_2]) &= [\sigma_1 - x(r_1), r_1] \diamond [\sigma_2 - x(r_2), r_2] && (68) \\ &= [\sigma_1 + \sigma_2 - x(r_1) - x(r_2) + \tilde{\delta}(r_1, r_2), r_1 r_2]. && (69) \end{aligned}$$

On the other hand, it can be readily verified that

$$\begin{aligned} \varphi([\sigma_1, r_1 \diamond [\sigma_2, r_2]) &= \varphi([\sigma_1 + \sigma_2 + \delta(r_1, r_2), r_1 r_2]) \\ &= [\sigma_1 + \sigma_2 + \delta(r_1, r_2) - x(r_1 r_2), r_1 r_2]. \quad (70) \end{aligned}$$

Thus, the equality of (69) and (70) follows directly from the equivalence of the local exponents. □

For completeness, we mention another general result concerning local isomorphisms. If z is a real nonzero constant and the local groups L and L' are constructed for δ and $\delta' = z\delta$, respectively, then the mapping $f : L \to L'$ such that $f([\sigma, r]) = [\sigma' = z\sigma, r' = r]$ defines an isomorphism between L and L'. In fact, $f([\sigma_1, r_1]) \diamond f([\sigma_2, r_2]) = [z\sigma_1, r_1] \diamond [z\sigma_2, r_2] = [z\sigma_1 + z\sigma_2 + \delta'(r_1, r_2), r_1 r_2]$. By its turn $f([\sigma_1, r_1] \diamond [\sigma_2, r_2]) = f([\sigma_1 + \sigma_2 + \delta(r_1, r_2), r_1 r_2]) = [z(\sigma_1 + \sigma_2 + \delta(r_1, r_2)), r_1 r_2]$ and, as $\delta' = z\delta$, $L \simeq L'$ locally.

Now, we can resume the discussion around Equation (64). As Lemma 3 asserts, studying isomorphism between local groups is a way to study local exponents. Moreover, Lemma 2 ensures that locally $L/C \simeq H$, that is L is the extension of H by C. Taken together, these two results point to the center of L, the C group, as the really relevant group to consider for the study of local exponents per se. The Lie algebra of the group L is at the heart of the question of whether a given representation is projective or genuine [15]. With effect, local exponents are (locally) equivalent to zero provided that a rearrangement of algebra generators removes their central counterpart. The study of continuity allows us to consider this result here from the group perspective: it is clear from the formulation that removing local exponents is possible if the group C whose elements are given by $[\sigma, e]$ can be ruled out from the analysis. This observation ultimately leads to the famous Bargmann's theorem, which states that central charges of semi-simple Lie algebras can always be excluded by redefining the algebra generators [4].

4. Final Remarks

We have discussed some aspects of Wigner's approach to dealing with (irreducible) representations of the Poincaré group. All the perspectives presented here focused on the rigor that Wigner devoted to the concept of representation's continuity itself. Indeed, this formal aspect has of course profound implications for the foundations of representation theory and, as we tried to make clear in the manuscript, its consequences are also relevant to physics.

In revisiting the underlying concepts for constructing the representation, we made a special effort to evoke all the relevant steps along with the proofs, and to provide a path through the concepts whose fulcrum rested on the matter of continuity. In this way, we were able to connect in advance (algebraic and topological) aspects of representation theory in physics, as well as some discussion about the projective or genuine representations issue, to the primitive idea of selecting admissible representatives.

At the risk of exaggerating somewhat, we would say that the formal consideration of some aspects of continuous representation in Hilbert spaces is perhaps one of the main ingredients of Wigner's success in finding a formal characterization of particles, along with other equally relevant aspects in the formulation, such as the exploration of induced representations [16,17] and finding of new representations.

The rigorous effort of Wigner's approach to formalize aspects previously bypassed by others is better explained by a well-known comment due to Wigner himself [18]:

> *The mathematical formulation of the physicist's often crude experience leads in an uncanny number of cases to an amazingly accurate description of a large class of phenomena. This shows that the mathematical language has more to commend it than being the only language which we can speak; it shows that it is, in a very real sense, the correct language.*

Just over 80 years have passed since the work celebrated here and, as befits a well-made work that survives the scrutiny of time, its fundamental concepts and approaches are still precise, elegant, and capable of revealing insightful perspectives.

Author Contributions: Conceptualization, J.M.H.d.S., G.M.C.d.R.; methodology, J.M.H.d.S. and G.M.C.d.R.; software, J.M.H.d.S. and G.M.C.d.R.; validation, J.M.H.d.S. and G.M.C.d.R.; formal analysis, J.M.H.d.S. and G.M.C.d.R.; investigation, J.M.H.d.S. and G.M.C.d.R.; resources, J.M.H.d.S. and G.M.C.d.R.; data curation, J.M.H.d.S. and G.M.C.d.R.; writing—original draft preparation, J.M.H.d.S. and G.M.C.d.R.; writing—review and editing, J.M.H.d.S. and G.M.C.d.R.; visualization, J.M.H.d.S. and G.M.C.d.R.; supervision, J.M.H.d.S. and G.M.C.d.R.; project administration, J.M.H.d.S. and G.M.C.d.R.; funding acquisition, J.M.H.d.S. and G.M.C.d.R. All authors have read and agreed to the published version of the manuscript.

Funding: J. M. Hoff da Silva was partially funded by Conselho Nacional de Desenvolvimento Científico e Tecnológico CNPq-Brazil. G. M. Caires da Rocha was funded by Coordenação de Aperfeiçoamento de Pessoal de Nível Superior—CAPES, Brazil.

Acknowledgments: JMHS thanks to CNPq (grant No. 303561/2018-1) for financial support.

Conflicts of Interest: The authors declare no conflict of interest. The funders had no role in the design of the study; in the collection, analyses, or interpretation of data; in the writing of the manuscript, or in the decision to publish the results.

Appendix A. Symmetry Representations

For completeness, we present here a step-by-step proof of the famous Wigner theorem of 1931 on symmetry operations in Hilbert space [1]. More often than never, the proof of this theorem is revisited in several contexts and with different levels of sophistication (see in [19] for a complete list of references and two quite interesting different proofs). Bargmann presented an elegant version of the proof [20], while a deep understanding of the Wigner viewpoints, together with some discussion about other proofs, is given in [21]. The presentation given here follows the steps performed by Weinberg [12] due to its completeness. After that, we also prove that for an identity component subgroup, every action is unitary. Let us start by contextualizing this rather important result.

We say that two descriptions of a given quantum mechanical system are isomorphic if there is a one-to-one correspondence, $\mathbf{\Psi} \leftrightarrow \mathbf{\Psi}'$, between the rays describing the physical system preserving probabilities, i.e., $\mathbf{\Phi} \cdot \mathbf{\Psi} = \mathbf{\Phi}' \cdot \mathbf{\Psi}'$. This is the desired situation for the description of the quantum system in two inertial reference frames connected by Lorentz transformations, for instance. Moreover, transformations that preserve the ray internal product (and hence the probabilities) are called symmetry transformations. As is well known, the Wigner theorem to be invoked here shows that every isomorphic ray correspondence engenders a (also one-to-one) vector correspondence in Hilbert space as $\psi' = U\psi$ and the general properties of U are listed in the theorem statement.

Theorem A1. *Every symmetry transformation in the Hilbert space of physical states can be represented by an operator that is either linear and unitary or anti-linear and anti-unitary.*

Proof. Let $\mathfrak{R}_1 \supset \psi_1$ and $\mathfrak{R}_2 \supset \psi_2$ be two rays in the Hilbert space $\mathcal{H}$ and, analogously, let $\mathfrak{R}'_1 \supset \psi'_1$ and $\mathfrak{R}'_2 \supset \psi'_2$ for transformed rays (and corresponding elements). Let $\varphi \in End(\mathcal{H})$ transforming $\mathfrak{R}$ into $\mathfrak{R}'$ be a symmetry transformation such that $|(\psi_1, \psi_2)|^2 = |(\psi'_1, \psi'_2)|^2$. In addition, we will also require the existence of φ^{-1} as a symmetry transformation. Denoting by $\{\psi_k\} \in \mathfrak{R}$ a complete orthonormal set of states, it is easy to see that orthonormality is inherited for $\{\psi'_k\} \in \mathfrak{R}'$, obtained by means of φ. Indeed, since $|(\psi_m, \psi_n)|^2 = \delta_{mn}$, we have $|(\psi'_m, \psi'_n)|^2 = \delta_{mn}$, from which

$$(\psi'_m, \psi'_n)(\psi'_m, \psi'_n)^* = \delta_{mn}. \tag{A1}$$

In this context, we note that although a notation tailored to the results for finite dimensional spaces has been used, the adaptation to general cases is straightforward. In accordance with more precise aspects (as completeness), the validation steps will be performed with no concern to a particularization for finite dimensions.

Let us consider Equation (A1) for the $m = n$ case. Let k be such that $(\psi'_m, \psi'_m) = k$. As it is well known (a quantum mechanical postulate) $0 \leq (\psi'_m, \psi'_m) \in \mathbb{R}$ (the equality holding for the null state). Therefore, Equation (A1) implies $k^2 = 1$, from which $(\psi'_m, \psi'_m) = 1$ necessarily. Consider now the case $m \neq n$ and take $(\psi'_m, \psi'_n) = a + ib$, where $a, b \in \mathbb{R}$. In this case Equation (A1) translates to $a^2 + b^2 = 0$, leading to $a = 0 = b$ and thus $(\psi'_m, \psi'_n) = 0$ for $n \neq m$. Therefore, $(\psi'_m, \psi'_n) = \delta_{mn}$ and the transformation φ does preserve orthonormality.

In accounting for completeness, suppose $\{\psi'_k\}$ is not a complete set. Then, there must exist a given state, say $\tilde{\psi}'$ which has not projection in the set (or a subset of) $\{\psi'_k\}$, that is $(\tilde{\psi}', \psi'_k) = 0$, $\forall k$. This fact would imply $|(\tilde{\psi}', \psi'_k)|^2 = 0$ and, via φ^{-1}, this would lead to $|(\tilde{\psi}', \psi'_k)|^2 = |(\tilde{\psi}, \psi_k)| = 0$, a clear contradiction as $\{\psi_k\}$ is complete. Therefore, the set $\{\psi'_k\}$ is also complete (and orthonormal) and a given basis of the original set is transformed into a basis in the arriving set.

Take now an element of the set $\{\psi_k\}$, say $\psi_{\bar{k}}$, and define $\phi_k = \frac{1}{\sqrt{2}}(\psi_{\bar{k}} + \psi_k)$ for $k \neq \bar{k}$. The element ϕ_k belongs to $\mathfrak{R}$. Notice, from the previous definition, that

$$|(\psi_{\bar{k}}, \phi_k)|^2 = \frac{1}{2}\left|(\psi_{\bar{k}}, \psi_{\bar{k}}) + (\psi_{\bar{k}}, \psi_k)\right|^2 = \frac{1}{2}. \tag{A2}$$

Besides, as we showed the transformation to preserve completeness, any vector belonging to $\mathfrak{R}'$ may be written as

$$\phi'_k = \sum_n \alpha_{kn} \psi'_n, \tag{A3}$$

from which we have

$$|(\psi'_{\bar{k}}, \phi'_k)|^2 = \left|\sum_n \alpha_{kn} (\psi'_{\bar{k}}, \psi'_n)\right|^2 = |\alpha_{k\bar{k}}|^2. \tag{A4}$$

For a symmetry transformation, Equations (A2) and (A4) should be equal, leading to $|\alpha_{k\bar{k}}| = \frac{1}{\sqrt{2}}$. This reasoning may be repeated for $|(\psi_k, \phi_k)|^2$ and $|(\psi'_k, \phi'_k)|^2$ resulting in $|\alpha_{kk}| = \frac{1}{\sqrt{2}}$. Furthermore, again repeating the same procedure, a bit of simple algebra shows that the coefficients α_{km} all vanish for $k \neq \bar{k}$ and $k \neq m$ (with $m \neq \bar{k}$). The first part of the demonstration is completed by choosing[8] $\alpha_{k\bar{k}} = 1/\sqrt{2} = \alpha_{kk}$ and calling the transformation engendering it by U, in such a way that

$$\phi'_k = U\phi_k = \frac{1}{\sqrt{2}}(U\psi_{\bar{k}} + U\psi_k). \tag{A5}$$

The task now is to extend the founded symmetry transformation to all Hilbert space. Let us now take $\psi = \sum_m \beta_m \psi_m$ from which we have

$$|(\psi_k, \psi)|^2 = \left|\sum_m \beta_m (\psi_k, \psi_m)\right|^2 = |\beta_k|^2. \tag{A6}$$

For a U transformed vector $\psi' = \sum_m \beta'_m U\psi_m$ it can be readily verified that $|(\psi'_k, \psi')|^2 = |\beta'_k|^2$ and, therefore, $|\beta_k|^2 = |\beta'_k|^2 \ \forall\ k$. In particular, $|\beta_{\bar{k}}|^2 = |\beta'_{\bar{k}}|^2$ and thus $|\beta_k|^2/|\beta_{\bar{k}}|^2 = |\beta'_k|^2/|\beta'_{\bar{k}}|^2$ or simply[9]

$$\left|\frac{\beta_k}{\beta_{\bar{k}}}\right|^2 = \left|\frac{\beta'_k}{\beta'_{\bar{k}}}\right|^2. \tag{A7}$$

Taking into account ϕ_k and ψ we have

$$|(\phi_k, \psi)|^2 = \left|\left(\frac{1}{\sqrt{2}}[\psi_{\bar{k}} + \psi_k], \sum_m \beta_m \psi_m\right)\right|^2 = \frac{1}{2}|\beta_{\bar{k}} + \beta_k|^2. \tag{A8}$$

Analogously

$$|(\phi'_k, \psi')|^2 = \left|\left(\frac{1}{\sqrt{2}}[U\psi_{\bar{k}} + U\psi_k], \sum_m \beta'_m U\psi_m\right)\right|^2 = \frac{1}{2}|\beta'_{\bar{k}} + \beta'_k|^2, \tag{A9}$$

allowing one to write $|\beta_{\bar{k}} + \beta_k|^2 = |\beta'_{\bar{k}} + \beta'_k|^2$. This last relation implies $|\beta'_{\bar{k}} + \beta'_k|^2/|\beta'_{\bar{k}}|^2 = |\beta_{\bar{k}} + \beta_k|^2/|\beta'_{\bar{k}}|^2$ and as $|\beta_1|^2 = |\beta'_1|^2$ we have $|\beta'_{\bar{k}} + \beta'_k|^2/|\beta'_{\bar{k}}|^2 = |\beta_{\bar{k}} + \beta_k|^2/|\beta_{\bar{k}}|^2$, or equivalently

$$\left|1 + \frac{\beta'_k}{\beta'_{\bar{k}}}\right|^2 = \left|1 + \frac{\beta_k}{\beta_{\bar{k}}}\right|^2. \tag{A10}$$

Calling for a moment $\beta_k/\beta_{\bar{k}} = x + iy$ (and similarly $\beta'_k/\beta'_{\bar{k}} = x' + iy'$), Equations (A7) and (A10) provide, respectively,

$$x^2 + y^2 = x'^2 + y'^2 \tag{A11}$$

and

$$(1 + x)^2 + y^2 = (1 + x')^2 + y'^2 \tag{A12}$$

whose combination demands $x = x'$ and $y = \pm y'$, that is,

$$Re\left(\frac{\beta'_k}{\beta'_{\bar{k}}}\right) = Re\left(\frac{\beta_k}{\beta_{\bar{k}}}\right),$$
$$Im\left(\frac{\beta'_k}{\beta'_{\bar{k}}}\right) = \pm Im\left(\frac{\beta_k}{\beta_{\bar{k}}}\right). \tag{A13}$$

These are the constraints to be taken in extending the U operation to all Hilbert space. By choosing the upper sign in Equation (A13) we have $\beta'_k/\beta'_{\bar{k}} = \beta_k/\beta_{\bar{k}}$, while the down sign implies $\beta'_k/\beta'_{\bar{k}} = \beta^*_k/\beta^*_{\bar{k}}$.

An important complement, due to Weinberg, to the original proof is the demonstration that the choice $\beta'_k/\beta'_{\bar{k}} = \beta_k/\beta_{\bar{k}}$ or $\beta'_k/\beta'_{\bar{k}} = \beta^*_k/\beta^*_{\bar{k}}$ must be made for all cases. To see this is indeed the case, suppose that for some k we have $\beta'_k/\beta'_{\bar{k}} = \beta_k/\beta_{\bar{k}}$, while for $m \neq k$ we have instead $\beta'^*_m/\beta'^*_{\bar{k}} = \beta_m/\beta_{\bar{k}}$. Define now a normalized χ such that

$$\chi = \frac{1}{\sqrt{3}}(\psi_{\bar{k}} + \psi_k + \psi_m), \tag{A14}$$

for which, of course, $\chi' = U\chi$. Repeating again the previous procedure being used so far, it is not difficult to see that

$$|(\chi,\psi)|^2 = \left|\left(\frac{1}{\sqrt{3}}(\psi_{\bar{k}} + \psi_k + \psi_m), \sum_n \beta_n\psi_n\right)\right|^2 = \frac{1}{3}|\beta_{\bar{k}} + \beta_k + \beta_m|^2, \tag{A15}$$

while $|(\chi',\psi')|^2 = |\beta'_{\bar{k}} + \beta'_k + \beta'_m|^2/3$, leading to $|\beta_{\bar{k}} + \beta_k + \beta_m|^2 = |\beta'_{\bar{k}} + \beta'_k + \beta'_m|^2$. Recalling that $|\beta_k|^2 = |\beta'_k|^2, \forall\, k$, we have

$$\frac{|\beta_{\bar{k}} + \beta_k + \beta_m|^2}{|\beta_{\bar{k}}|^2} = \frac{|\beta'_{\bar{k}} + \beta'_k + \beta'_m|^2}{|\beta'_{\bar{k}}|^2}. \tag{A16}$$

Taking into account the supposition assumed within this paragraph, Equation (A16) may be recast in the form

$$\left|1 + \frac{\beta_k}{\beta_{\bar{k}}} + \frac{\beta^*_m}{\beta^*_{\bar{k}}}\right|^2 = \left|1 + \frac{\beta_k}{\beta_{\bar{k}}} + \frac{\beta_m}{\beta_{\bar{k}}}\right|^2. \tag{A17}$$

A bit of algebra shows that the constraint presented in Equation (A17) translates to

$$\frac{\beta_m\beta_k}{\beta_{\bar{k}}\beta_{\bar{k}}} + \frac{\beta^*_m\beta^*_k}{\beta^*_{\bar{k}}\beta^*_{\bar{k}}} = \frac{\beta^*_m\beta_k}{\beta^*_{\bar{k}}\beta_{\bar{k}}} + \frac{\beta_m\beta^*_k}{\beta_{\bar{k}}\beta^*_{\bar{k}}}, \tag{A18}$$

or simply

$$Re\left(\frac{\beta_k\beta^*_m}{\beta_{\bar{k}}\beta^*_{\bar{k}}}\right) = Re\left(\frac{\beta_k\beta_m}{\beta_{\bar{k}}\beta_{\bar{k}}}\right). \tag{A19}$$

Finally, calling $\beta_k/\beta_{\bar{k}} \equiv u = u_1 + iu_2$ and $\beta_m/\beta_{\bar{k}} \equiv v = v_1 + iv_2$, the above equation $(Re(uv^*) = Re(uv))$ can only be satisfied if $u_2v_2 = 0$, that is

$$Im\left(\frac{\beta_k}{\beta_{\bar{k}}}\right)Im\left(\frac{\beta_m}{\beta_{\bar{k}}}\right) = 0, \tag{A20}$$

a clear contradiction regarding a complex space of states. Therefore, one is left with either $\beta'_k/\beta'_{\bar{k}} = \beta_k/\beta_{\bar{k}}$ or $\beta'_k/\beta'_{\bar{k}} = \beta^*_k/\beta^*_{\bar{k}}$. In the case $\beta'_k = \beta_k, \forall\, k$, then

$$U\Big(\sum_k \beta_k\psi_k\Big) = \sum_k \beta_k U\psi_k, \tag{A21}$$

or else $(\beta'_k = \beta^*_k, \forall\, k)$

$$U\Big(\sum_k \beta_k\psi_k\Big) = \sum_k \beta^*_k U\psi_k. \tag{A22}$$

It is important to have a clear-cut of what was demonstrated. Equations (A21) and (A22) show that a same choice must necessarily be made for element inside a given vector state. The theorem demonstration is finalized by showing that it is impossible for a given same transformation that a given vector state transform like (A21) and others as (A22) dictates. Taking advantage of the finite dimension spaces notation, let $\mathbb{V} \subset \mathcal{H}$ be a Hilbert subspace spanned by $\{\psi_k\}$ and $k = 1,\ldots,l,\ldots,\dim\mathbb{V}$. Suppose the existence of φ for which $\varphi' = \sum_k a_k U\psi_k$ and η such that $\eta' = \sum_k b^*_k U\psi_k$ for the same operator U. Within this context, a vector $\rho = \sum_{k=1}^{l-1} a_k\psi_k + \sum_{k=l}^{\dim\mathbb{V}} b_k\psi_k$ would transform under U so as to violate the result encoded in (A21) and (A22). The finalization of the proof now follows straightforwardly. Let $\psi_1 = \sum_k \alpha_k \psi_k$ and $\psi_2 = \sum_k \beta_k \psi_k$. Assuming U acting according to (A21), i.e., linearly, we have

$$(U\psi_1, U\psi_2) = \sum_{m,n} \alpha_m^* \beta_n (U\psi_m, U\psi_n) = \sum_{m,n} \alpha_m^* \beta_n \delta_{mn} = (\psi_1, \psi_2), \tag{A23}$$

asserting U as unitary. By instead demanding (A22), its anti-linearity implies

$$(U\psi_1, U\psi_2) = \sum_{m,n} \alpha_m \beta_n^* (U\psi_m, U\psi_n) = \sum_{m,n} \alpha_m \beta_n^* \delta_{mn} = (\psi_2, \psi_1) = (\psi_1, \psi_2)^*, \tag{A24}$$

from which an anti-unitary action may be read. □

We finalize this appendix by recalling that for the largest connected subgroup (the identity subgroup), the orthochronous proper Lorentz subgroup, for example, every operator ray is unitary.

Proposition A1. *Let H be a connected group (or equivalently an identity*[10] *(sub)group). Then, $\mathbf{U}_r$ is unitary for all $r \in H$.*

Proof. In a suitable neighborhood $\mathfrak{N} \in H$ containing e, every group element r can be written as $r = s^2$ (s being, of course, another group element). By the theorem just exposed, operators acting upon the Hilbert space $\mathcal{H}$ are either linear and unitary or anti-linear and anti-unitary, and thus an isomorphic ray correspondence defines ray operators endowed with the same properties. It turns out, however, that the square of a unitary or anti-unitary operator is unitary. To see that, recall that if $\mathbf{U}_s$ is anti-linear, then $\mathbf{U}_r = \mathbf{U}_s^2$ and for any $\phi, \psi \in \mathcal{H}$ it follows

$$(\mathbf{U}_r\phi, \mathbf{U}_r\psi) = (\mathbf{U}_s[\mathbf{U}_s\phi], \mathbf{U}_s[\mathbf{U}_s\psi]) = (\mathbf{U}_s\psi, \mathbf{U}_s\phi) = (\phi, \psi). \tag{A25}$$

In the case of $\mathbf{U}_s$ the proof follows straightforwardly.[11]

As H is connected, every element $r \in H$ equals a given finite product, say $r = \prod_{i=1}^n r_i$, of elements in $\mathfrak{N}$. Hence $\mathbf{U}_r = \prod_{i=1}^n \mathbf{U}_{r_i}$. Now, for every $\mathbf{U}_{r_i}$ take a partition, as before, in such a way that

$$\mathbf{U}_r = \prod_{i=1}^n \mathbf{U}_{s_i}^2. \tag{A26}$$

Then, the ray operator $\mathbf{U}_r$ is given by the a finite product of unitary (ray) operators, from which it is seen that $\mathbf{U}_r$ is itself unitary. □

Notes

1 This choice was made by Wigner [3], p. 169.

2 In fact, calling $(\psi, \phi) = x + iy$ with $x, y \in \mathbb{R}$, we see that $[1 - Re(\psi, \phi)]_{\max} = (1 - x)_{\max} \leq [(1 - x)^2_{\max} + y^2]^{1/2}$ from which the inequality follows.

3 The reader may here appreciate the inventiveness of Wigner's approach: in general considerations about Quantum Mechanics the dimension of the underling Hilbert space is not *a priori* specified. It is indeed so, as it is the quantum mechanical problem that dictates the dimension. This simple observation reveals the finesse of Wigner's procedure in proving the continuity for ϕ and χ.

4 The rest of the proof follows strictly the remaining steps of the proof performed in the preceding section.

5 In fact, for a complex A, $(AU_m\psi, AU_m\psi)^{1/2} = |A|(U_m\psi, U_m\psi)^{1/2} = |A|(\psi, U_{m^{-1}}U_m\psi)^{1/2} = |A|$ for normalized vectors.

6 Technically additional care should be take in the neighborhood's definition. In fact, with $\mathfrak{N}$ and $\mathfrak{N}'$ being the neighborhoods of e where U_r' and U_r are, respectively, defined, then the relation $U_r' = \phi(r)U_r$ takes place in a neighborhood $\mathfrak{K}$ of e such that $\mathfrak{K} \subset (\mathfrak{N} \cap \mathfrak{N}')$. Besides, for admissible representative sets, Equation (45) is valid for $\phi(r)$ defined in $\mathfrak{K}$ also encompassing products as rs.

7 It should be mentioned that in order to extrapolate local results for the whole group it is also necessary the group to be compact. This characteristic is, however, fulfilled by every Lie group [13].

8 This choice, as it should be clear in the course of the proof, does not change the conclusions.

9 In the case that β_1 is null, one should particularize another k and perform the same analysis. Since the ψ state is assumed to exists, eventually a given k would lead to a non vanishing coefficient.

0 A technicality should be mentioned here: in general, an identity component refers to the largest connected set in the manifold associated to the group containing the identity element e of the group.

1 Note, in addition, that from the isometry condition $\mathbf{U}_r^{\dagger}\mathbf{U}_r = 1$ (and the coisometry condition $\mathbf{U}_r\mathbf{U}_r^{\dagger} = 1$ as well) this property is readily obtained.

References

1. Wigner, E.P. *Gruppentheorie und ihre Anwendung auf die Quantenmechanik der Atomspektren*; F. Vieweg und Sohn: Braunschweig, Germany, 1931; (English Translation, Academic Press: New York, NY, USA, 1959).
2. Weyl, H. *Gruppentheorie und Quantenmechanik*; Z. Physik: Leipzig, Germany, 1928; (English Edition, Dover (1950)).
3. Wigner, E.P. On Unitary Representations of the Inhomogeneous Lorentz Group. *Ann. Math.* **1939**, *40*, 149. [CrossRef]
4. Bargmann, V. On Unitary Ray Representations of Continuous Groups. *Ann. Math.* **1954**, *59*, 1. [CrossRef]
5. Majorana, E. Teoria Relativistica di Particelle Con Momento Intrinseco Arbitrario. *Nuovo Cim.* **1932**, *9*, 335. [CrossRef]
6. Dirac, P.A.M. Relativistic wave equations. *Proc. R. Soc. A* **1936**, *155*, 447.
7. Neeb, K.-H. Positive energy representations and continuity of projective representations for general topological groups. *Glasgow Math. J.* **2014**, *56*, 295. [CrossRef]
8. Janssens, B.; Neeb, K.-H. Projective unitary representations of infinite dimensional Lie groups. *Kyoto J. Math.* **2019**, *59*, 293. [CrossRef]
9. Wawrzycki, J. A Generalization of the Bargmann's Theory of Ray Representations. *Commun. Math. Phys.* **2004**, *250*, 215. [CrossRef]
10. Filippeli, G. Time Dependent Quantum Generators for the Galilei Group. *J. Math. Phys.* **2011**, *52*, 083510. [CrossRef]
11. Nakahara, M. *Geometry, Topology and Physics*; Taylor & Francis: New York, NY, USA, 2003.
12. Weinberg, S. Foundations. In *The Quantum Theory of Fields*; Cambridge University Press: New York, NY, USA, 2005; Volume I.
13. Hall, B. *Lie Groups, Lie Algebras, and Representations—An Elementary Introduction*; Springer: Berlin, Germany, 2015.
14. Nash, C.; Sen, S. *Topology and Geometry for Physicists*; Academic Press: London, UK, 1983.
15. Gilmore, R. *Lie Groups, Lie Algebras and Some of Their Applications*; Dover: New York, NY, USA, 2005.
16. Mackey, G.W. Induced Representations of Locally Compact Groups I. *Ann. Math.* **1952**, *55*, 101. [CrossRef]
17. Mackey, G.W. Induced Representations of Locally Compact Groups II. The Frobenius Reciprocity Theorem. *Ann. Math.* **1953**, *58*, 193. [CrossRef]
18. Wigner, E.P. The Unreasonable Effectiveness of Mathematics in the Natural Sciences. *Commun. Pure Appl. Math.* **1960**, *13*, 1. [CrossRef]
19. Simon, R.; Mukunda, N.; Chaturved, S.; Srinivasan, V. Two elementary proofs of the Wigner theorem on symmetry in quantum mechanics. *Phys. Lett. A* **2008**, *372*, 6847–6852 [CrossRef]
20. Bargmann, V. Note on Wigner's theorem in quantum mechanics. *J. Math. Phys.* **1964**, *5* 862.
21. Chevalier, G. Wigner's Theorem and its Generalizations. In *Handbook of Quantum Logic and Quantum Structures: Quantum Structures*; Engesser, K., Gabbay, D.M., Lehmann, D., Eds.; Elsevier: Amsterdam, The Netherlands, 2007; Volume 429.

universe
MDPI

Review

Fundamental Symmetries and Spacetime Geometries in Gauge Theories of Gravity—Prospects for Unified Field Theories

Francisco Cabral [1,*], Francisco S. N. Lobo [1] and Diego Rubiera-Garcia [2]

1 Instituto de Astrofísica e Ciências do Espaço, Faculdade de Ciências da Universidade de Lisboa, Edifício C8, Campo Grande, P-1749-016 Lisbon, Portugal; fslobo@fc.ul.pt
2 Departamento de Física Teórica and IPARCOS, Universidad Complutense de Madrid, E-28040 Madrid, Spain; drubiera@ucm.es
* Correspondence: ftcabral@fc.ul.pt

Received: 13 November 2020; Accepted: 7 December 2020; Published: 11 December 2020

Abstract: Gravity can be formulated as a gauge theory by combining symmetry principles and geometrical methods in a consistent mathematical framework. The gauge approach to gravity leads directly to non-Euclidean, post-Riemannian spacetime geometries, providing the adequate formalism for metric-affine theories of gravity with curvature, torsion and non-metricity. In this paper, we analyze the structure of gauge theories of gravity and consider the relation between fundamental geometrical objects and symmetry principles as well as different spacetime paradigms. Special attention is given to Poincaré gauge theories of gravity, their field equations and Noether conserved currents, which are the sources of gravity. We then discuss several topics of the gauge approach to gravitational phenomena, namely, quadratic Poincaré gauge models, the Einstein-Cartan-Sciama-Kibble theory, the teleparallel equivalent of general relativity, quadratic metric-affine Lagrangians, non-Lorentzian connections, and the breaking of Lorentz invariance in the presence of non-metricity. We also highlight the probing of post-Riemannian geometries with test matter. Finally, we briefly discuss some perspectives regarding the role of both geometrical methods and symmetry principles towards unified field theories and a new spacetime paradigm, motivated from the gauge approach to gravity.

Keywords: gauge field theory; Yang-Mills fields; modified gravity; non-Riemannian geometry; spacetime symmetries

1. Introduction

The success of Einstein's General Theory of Relativity (GR) to describe the behaviour of the gravitational interaction continues to amaze us. It has passed all tests performed so far: Solar System observations and binary pulsars [1], stellar orbits around the central galactic black hole [2], gravitational waves (GWs) from coalescing compact objects (black holes and neutron stars) [3–6], or the indirect observation of the black hole horizon with the Event Horizon Telescope [7], among others. At the same time, it provides us with the observationally valid framework for the standard cosmological paradigm when supplemented with the (cold) dark matter and dark energy hypothesis [8].

Soon after Einstein formulated GR in its final form, Weyl introduced the notion of gauge transformations in an attempt to unify gravity and electromagnetism [9]. By extending the local Lorentz group to include scale transformations (dilatations), he was led to assume what we call nowadays a Riemann-Weyl spacetime geometry, namely, a post-Riemann geometry with (the trace-vector part of) non-metricity in addition to the familiar curvature of GR. Later on, it was discovered that the electromagnetic field itself is intimately related to local internal symmetries, under the $U(1)$ group that acts on the four-spinor fields of charged matter [10]. In the 1950s, Yang and Mills [11] further explored the notion of gauge symmetries in field theories going beyond the $U(1)$ group to include non-abelian Lie groups ($SU(2)$), in order to address nuclear physics, while Utiyama [12] extended the gauge principle to all semi-simple Lie groups including the Lorentz group.

The gauge principle is based on the localization of the rigid global symmetry group of a field theory, introducing a new interaction described by the gauge potential. The latter is a compensating field that makes it possible for the matter Lagrangian to be locally invariant under the symmetry group and is included in the covariant derivative of the theory. There is a clear geometrical interpretation of the gauge potential as the connection of the fiber bundle, which is the manifold obtained from the base spacetime manifold and the set of all fibers. These are attached at each spacetime point and are the (vector, tensor or spinor) spaces of representation of the local symmetries. In the geometrical interpretation, the imposition of local symmetries implies that the geometry of the fiber bundle is non-Euclidean, and the gauge field strengths are the curvatures of such a manifold.

The gauge formulation of gravity was resumed through the works of Sciama [13] and Kibble [14], who gauged the (rigid) Poincaré group of Minkowski spacetime symmetries. They arrived at what is now known as a Riemann-Cartan (RC) geometry, and to the corresponding Einstein-Cartan-Sciama-Kibble (ECSK) gravity, with non-vanishing torsion and curvature. This is a natural extension of GR, which is able to successfully incorporate the intrinsic spin of fermions as a source of gravity, while passing all weak-field limit tests. Moreover, the theory has no free parameters but rather a single new scale given by Cartan's density, which yields many relevant applications in cosmology and astrophysics [15–23]. It is worth noting that ECSK is the simplest of all Poincaré gauge theories of gravity (PGTG). Beyond the Poincaré group we have, for instance, the Weyl and the conformal groups, which live on a subset of a general metric-affine geometry, with non-vanishing curvature, torsion and non-metricity (for a detailed analysis and reviews on several topics of the gauge approach to gravity see the remarkable works in References [24–26]). By extending the gauge symmetry group of gravity, one is naturally led to extend the spacetime geometry paradigm as well. The geometrical methods and symmetry principles, motivated by gauge theories of gravity, together with post-Riemannian geometries and Yang-Mills gauge fields, are expected to play a key role towards a unified field theory. The corresponding physical description might lead to a new spacetime paradigm extending the notion of the classical spacetime manifold where physical fields propagate on it and affect its geometry. Instead, it is plausible to expect a consistent unified physical manifold, where the properties of spacetime, matter fields and vacuum are manifestations of the same (unified) fundamental physical reality.

The main aim of this work is to review and discuss the basics and recent developments within gauge theories of gravity and post-Riemann geometries in connection to their perspectives for achieving an unified field theory. This is a vital part of the effort to understand the nature of spacetime and gravity in those regimes where the standard picture provided by GR may break down, challenging our current ideas on the spacetime paradigm. This article complements our previous works: an extension of the Einstein-Cartan-Dirac theory where an electromagnetic (Maxwell) contribution minimally coupled to torsion is introduced (which breaks the $U(1)$ gauge symmetry [27]), and its extension to analyze in detail the physics of this model with Dirac and Maxwell fields minimally coupled to the spacetime torsion [28,29].

This work is organized as follows: in Section 2 we introduce the fundamental geometrical objects of the spacetime manifold in the formalism of exterior forms and its relation to spacetime symmetries. In Section 3 we briefly summarize the metric-affine approach to gravity using these geometrical methods and symmetry principles, and discuss several paradigms for the spacetime geometry, including the general metric-affine geometry. In Section 4, we outline the general structure of the gauge approach to gravity. We illustrate it with the case of PGTG, consider the field equations, the Noether conserved currents (sources of gravity), and include discussions on several topics in gauge theories of gravity, such as quadratic Poincaré gauge models, the ECSK theory, the teleparallel equivalent of GR (an example of a translational gauge model), quadratic metric-affine Lagrangians, the breaking of Lorentz invariance in the presence of non-metricity, the nature of the hypermomentum currents and the probing of post-Riemann geometries with test matter. Finally, in Section 5 we discuss some perspectives for unified field theories and a new spacetime paradigm motivated from the geometrical methods and symmetry principles of the gauge approach to gravity.

2. Spacetime Symmetries and Post-Riemann Geometries

2.1. Fundamental Geometrical Structures of Spacetime

Let us begin by considering a four-dimensional differential manifold $\mathcal{M}$ as an approximate representation of the physical spacetime, and introduce the fundamental geometrical objects and its relation to the group theory of spacetime symmetries. At each point P of the spacetime manifold we introduce the set of four linearly independent vectors $\{\bar{e}_0, \bar{e}_1, \bar{e}_2, \bar{e}_3\}$, with $\bar{e}_0 \equiv \partial_0$, $\bar{e}_1 \equiv \partial_1$, $\bar{e}_2 \equiv \partial_2$, $\bar{e}_3 \equiv \partial_3$, which constitute the vector coordinate (holonomic) basis, where each vector is tangent to a coordinate line. This is called a linear frame basis. Similarly, at the same point we introduce the dual co-frame basis $\{\bar{\theta}^0, \bar{\theta}^1, \bar{\theta}^2, \bar{\theta}^3\}$, with $\bar{\theta}^0 \equiv dx^0$, $\bar{\theta}^1 \equiv dx^1$, $\bar{\theta}^2 \equiv dx^2$, $\bar{\theta}^3 \equiv dx^3$, which satisfies $\bar{e}_b \lrcorner \bar{\theta}^a = \delta^a_b$[1]. Any of these basis can be called the natural (coordinate/holonomic) frame/co-frame. This geometrical structure comes naturally with the notion of coordinates on the spacetime manifold, as it is intrinsic to the coordinates structure.

Note that one can choose any set of linearly independent vectors/co-vectors to form arbitrary linear frames and co-frames. To this effect, we consider the independent combinations $e_b = e_b{}^\nu \partial_\nu$ and $\theta^a = \theta^a{}_\mu dx^\mu$. The indices $a, b = 0,1,2,3$ are called anholonomic indices, sometimes known as symmetry or group indices, and play a fundamental role in the gauge approach to gravity due to its connection to spacetime symmetries. It is clear that for the natural frame/co-frame we have $\bar{e}_b = \delta_b{}^\nu \partial_\nu$ and $\bar{\theta}^a = \delta^a{}_\mu dx^\mu$. For arbitrary (non-coordinate) anholonomic vector basis, the Lie brackets is non-vanishing, $[U, V] \equiv \pounds_U V \neq 0$, for any two vectors U, V in the basis. In relation to this one can show, using the definitions and duality relations already introduced, the following algebra $[e_a, e_b] = f_{ab}{}^c e_c$, where the objects $f_{ab}{}^c$ are typically known as the (group) structure constants. This algebra can be used to characterize the local spacetime symmetries of the tangent/cotangent spaces.

In four dimensions, the set of vector valued 1-forms θ^a constitute 16 independent components $\theta^a{}_\mu$ (the tetrads) and are the potentials for the group of (local) spacetime translations $T(4)$[2]. This group has four generators, which therefore entails four potentials and four field strengths $T^a = \frac{1}{2} T^a{}_{\mu\nu} dx^\mu \wedge dx^\nu$, the latter being a vector valued 2-form field, and corresponds to the torsion of the spacetime manifold. It is given by

$$T^a = D\theta^a = d\theta^a + \Gamma^a{}_b \wedge \theta^b \,, \tag{1}$$

where D is the (gauge) covariant exterior derivative, d is the exterior derivative, a kind of curl operator that raises the degree of any p-form, $\wedge$ is the wedge product[3] and in the second term on the right-hand

---

1 The symbol $\lrcorner$ stands for the interior product defined in differential topology and linear algebra, also called contraction operator, which gives a contraction between a p-form and a vector, resulting in a $(p-1)$-form.

2 This group is actually related to diffeomorphisms (and to our freedom to choose any system of coordinates without altering the physical description, no matter the theory of the gravitational field we are working with), playing the role of gauge transformations of gravity [30].

3 Accordingly, $d\theta^a$ is a 2-form given by $d\theta^a = \partial_{[\mu} \theta^a{}_{\nu]} dx^\mu \wedge dx^\nu$. Similarly, if v is a p form and w is a k-form, then $v \wedge w$ is a $(p+k)$-form with $v \wedge w = (-1)^{p \times k} w \wedge v$. The (gauge) covariant exterior derivative of a generic tensor valued p-form, denoted by $V^a{}_b$, used in this paper is given by $DV^a{}_b = dV^a{}_b + \Gamma^a{}_c \wedge V^c{}_b + (-1)^p \Gamma^c{}_b \wedge V^a{}_c$.

side, sometimes called the non-trivial part, is the linear connection 1-form $\Gamma^a{}_b$. The torsion 2-form has $4 \times 6 = 24$ independent components given explicitly by

$$T^a{}_{\mu\nu} = 2\partial_{[\mu}\theta^a{}_{\nu]} + 2\Gamma^a{}_{c[\mu}\theta^c{}_{\nu]} \tag{2}$$

where brackets denote antisymmetrization.

Let us now turn our attention to the linear (affine) connection $\Gamma^a{}_b$, which is a tensor valued 1-form that connects neighbouring points of the manifold. Accordingly, if $v = v^a e_a$ is a vector parallel-transported by an infinitesimal displacement δx, the difference between the parallely-displaced vector and the original vector is given by

$$v^a_{\|}(x+\delta x) - v^a(x) = \delta_{\|} v^a = -\Gamma^a{}_b v^b, \tag{3}$$

with $\Gamma^a{}_b = \Gamma^a{}_{b\mu}dx^\mu$. Therefore, upon an arbitrary infinitesimal displacement from a given point in the spacetime manifold, the linear frame is transformed (e.g., under a Lorentz rotation or some more generic linear transformation) and the connection gives a measure of such a change in the linear frame. It is worth emphasizing that the introduction of the affine structure of the spacetime manifold $\mathcal{M}$ is mandatory in order to be able to compare between tensor quantities on different spacetime points. Indeed, the affine connection allows for the definition of a covariant derivative, and in this way it establishes a rule for the parallel transport of tensors along curves of $\mathcal{M}$. Moreover, it allows to determine the affine geodesics (straightest lines) on $\mathcal{M}$, which do not necessarily coincide with extremal geodesics ("shortest" paths).

The affine connection is mathematically a very rich object. Among its many properties we underline the following two: (i) the difference of two connections is a tensor; (ii) under a transformation $\Gamma \to \Gamma + \boldsymbol{\o}$, where $\boldsymbol{\o}$ is a tensor, the covariant derivative of tensors retains its covariance. This means that the components of the covariant derivative of a tensor still transform as a tensor, which implies that one can incorporate new degrees of freedom in the geometrical (affine) structure of spacetime while preserving the covariance of the equations. Such extensions of the connection involve an extended spacetime geometry, while the inclusion of extra (gauge) degrees of freedom in a field theory requires extending the local symmetry group. These two facts are inevitably interconnected in gauge theories of gravity.

In four dimensions, the set of tensor valued 1-forms $\Gamma^a{}_b$ constitute 64 independent components $\Gamma^a{}_{b\mu}$ and are the potentials for the four-dimensional group of general linear transformations $GL(4,\Re)$. With the definition of the linear frame/coframe the arbitrary (general) non-degenerate linear transformations of spacetime coordinates can be defined and $\Gamma^a{}_b$ turn out to be the generators of such a group of transformations. It has 16 generators, therefore we have 16 potentials which are analogous to the Yang-Mills potentials of the $SU(3)$ group. The associated field strength $R^a{}_b$ is a tensor valued 2-form field $R^a{}_b = \frac{1}{2}R^a{}_{b\mu\nu}dx^\mu \wedge dx^\nu$, corresponding to the curvature of the spacetime manifold, which can be written explicitly as

$$R^a{}_b = d\Gamma^a{}_b + \Gamma^a{}_c \wedge \Gamma^c{}_b\,. \tag{4}$$

The curvature 2-form has $16 \times 6 = 96$ independent components, given by

$$R^a{}_{b\mu\nu} = 2\partial_{[\mu}\Gamma^a{}_{b|\nu]} + 2\Gamma^a{}_{c[\mu}\Gamma^c{}_{b|\nu]}\,. \tag{5}$$

The linear connection 1-form can be decomposed according to

$$\Gamma_{ab} = \tilde{\Gamma}_{ab} + N_{ab} = \tilde{\Gamma}_{ab} + \frac{1}{2}Q_{ab} + N_{[ab]}\,, \tag{6}$$

where the Levi-Civita part of the connection, $\tilde{\Gamma}_{ab}$, obeys the Cartan structure equation $d\theta^a + \tilde{\Gamma}^a{}_b \wedge \theta^b = 0$, and $N^a{}_b$ is the so-called distortion 1-form characterizing the post-Riemannian geometries. In particular, one finds that $Q_{ab} = 2N_{(ab)}$ and $T^a = N^a{}_b \wedge \theta^b$. If the linear connection obeys the condition $\Gamma^{ab} = -\Gamma^{ba}$, then it is called a Lorentzian connection (or spin connection) and corresponds to 24 independent components. This is the case when the linear connection is the potential for the Lorentz group $SO(1,3)$ (not the full linear group GL(4, $\Re$)). Accordingly, as we shall see later, the Lorentzian connection is the linear connection of PGTG.

Though one can argue that the spacetime metric is not as fundamental as the affine connection (and indeed a purely affine formulation of gravity is possible, see for example, Reference [31]), one can introduce it in order to measure time and space intervals as well as angles. Thus, let us define the Lorentzian metric as the $(0,2)$ tensor $g = g_{ab}\theta^a \otimes \theta^b$, where $a, b = 0,1,2,3$ are anholonomic indices, and the spacetime metric components in the coordinate frame are given by $g_{\mu\nu} = \theta^a{}_\mu \theta^a{}_\nu g_{ab}$. The Lorentzian metric g_{ab} is the metric of the tangent space required to compute the inner product between anholonomic vectors and make a map between these vectors and the corresponding dual co-vectors $v_b = g_{ab}v^a$. As for the spacetime metric $g_{\mu\nu}$, it establishes maps between the contravariant components of vectors and the covariant components of the corresponding dual covectors in the coordinate (holonomic basis), and defines the inner products $g(u,v) = g_{\alpha\beta}u^\alpha v^\beta = u_\alpha v^\beta$. Therefore, the spacetime metric can be seen as the deformation of the Lorentzian (tangent space) metric according to $g_{\mu\nu} = \Omega^{ab}{}_{\mu\nu} g_{ab}$, with $\Omega^{ab}{}_{\mu\nu} \equiv \theta^a{}_\mu \theta^a{}_\nu$ being a deformation tensor. Since the tangent space has a pseudo-Euclidean geometry, then if the spacetime manifold has non-Euclidean geometry the deformation tensor has to vary from point to point. From now on we consider geometries where the spacetime metric is symmetric $g_{\alpha\beta} = g_{\beta\alpha}$, non-degenerate, $g = \det(g_{\mu\nu}) \neq 0$, and determines the local line element $ds^2 = g_{\alpha\beta}dx^\alpha dx^\beta$ with a Lorentzian signature (± 2). As for the metric g_{ab}, it is possible to see it as a kind of potential whose corresponding field strength, $Q_{ab} = Dg_{ab} = dg_{ab} + \Gamma^c{}_a \wedge g_{cb} + \Gamma^c{}_b \wedge g_{ac} = 2\Gamma_{(ab)}$, is a tensor valued 1-form $Q_{ab} = Q_{ab\mu}dx^\mu$ with $10 \times 4 = 40$ independent components $Q_{ab\mu}$. This field strength corresponds to the non-metricity tensor valued 1-form and one concludes that, if the connection is non-Lorentzian ($\Gamma^{ab} \neq -\Gamma^{ba}$), then non-metricity is non-vanishing.

In (pseudo) Euclidean geometries such as that of Minkowski spacetime, there is always a coordinate system where the components of the connection (the Christoffel symbols) and their derivatives vanish, whereas in a (pseudo) Riemann geometry with non-vanishing curvature, one can find a local geodesic system of coordinates where the connection vanishes and the metric is given by the Minkowski metric (a freely falling frame), but the derivatives of the connection cannot be set to zero. In such geometries the so-called Levi-Civita connection and the metric are fundamentally related and thus not independent from each other. Indeed, in such a case the metricity condition (vanishing of the covariant derivatives of the metric) holds, implying that the connection is proportional to the first derivatives of the metric (recall that the Levi-Civita connection is the only symmetric connection that obeys the metricity condition) and as such, in GR, the presence of a physical gravitational field is traced to the non-vanishing of the second derivatives of the metric. The Weyl part of the Riemannian curvature is not absent in a freely falling frame, which translates into tidal effects. But both the symmetry of a connection and the metricity condition can be relaxed leading to more general geometries with torsion and non-metricity respectively, as we shall see next.

2.2. The Decomposition of the Affine Connection

We are now ready to bring here the well known result that any affine connection can be decomposed into three independent components. In holonomic (coordinate) components they are expressed as

$$\Gamma^\lambda{}_{\mu\nu} = \tilde{\Gamma}^\lambda{}_{\mu\nu} + K^\lambda{}_{\mu\nu} + L^\lambda{}_{\mu\nu}\,, \tag{7}$$

where $\tilde{\Gamma}^\lambda{}_{\mu\nu}$ is the Levi-Civita connection associated to the Riemannian curvature

$$\tilde{R}^\alpha_{\beta\mu\nu} = 2\partial_{[\mu}\tilde{\Gamma}^\alpha_{\nu]\beta} + 2\tilde{\Gamma}^\alpha_{[\mu|\lambda|}\tilde{\Gamma}^\lambda_{\nu]\beta}\,, \tag{8}$$

the term $K^{\lambda}{}_{\mu\nu}$ is associated to torsion, $T^{\lambda}_{\alpha\beta} \equiv \Gamma^{\lambda}_{[\alpha\beta]}$, and is denoted contortion:

$$K^{\lambda}{}_{\mu\nu} \equiv T^{\lambda}{}_{\mu\nu} - 2T_{(\mu}{}^{\lambda}{}_{\nu)}, \tag{9}$$

while the term $L^{\lambda}{}_{\mu\nu}$ is associated to non-metricity $Q_{\rho\mu\nu} \equiv \nabla_{\rho} g_{\mu\nu}$, and is called disformation:

$$L^{\lambda}{}_{\mu\nu} \equiv \frac{1}{2} g^{\lambda\beta} \left(-Q_{\mu\beta\nu} - Q_{\nu\beta\mu} + Q_{\beta\mu\nu} \right) . \tag{10}$$

This decomposition naturally highlights the metric-affine character of the most general spacetime geometry, made up of curvature, torsion and non-metricity. Let us briefly analyze each component separately.

2.2.1. Curvature

In an holonomic basis the curvature of a connection has the 96 independent components

$$R^{\alpha}{}_{\beta\mu\nu} = \partial_{\mu} \Gamma^{\alpha}{}_{\beta\nu} - \partial_{\nu} \Gamma^{\alpha}{}_{\beta\mu} + \Gamma^{\alpha}{}_{\lambda\mu} \Gamma^{\lambda}{}_{\beta\nu} - \Gamma^{\alpha}{}_{\lambda\nu} \Gamma^{\lambda}{}_{\beta\mu} . \tag{11}$$

Consider at some point P of the manifold the vectors $U = d/d\lambda$ and $V = d/d\sigma$, tangent to two curves intersecting at P, where λ and σ are some respective affine parameters labelling each curve. Since the curvature tensor is of $(1,3)$ type, it can be applied (contracted) to U and V and then to some other vector field Z, which yields the resulting vector field

$$R(U,V)Z = \nabla_U \nabla_V Z - \nabla_V \nabla_U Z - \nabla_{[U,V]} Z , \tag{12}$$

where $[U,V] = \pounds_U V$ represents the Lie derivative (of V with respect to U). To clarify its geometrical meaning, let us consider an infinitesimal closed loop, with $ds^{\mu\nu}$ being the surface element spanned by such a loop. After a parallel transport of some vector v along the loop, it is found that the initial and final vectors do not coincide: there is a rotation with a difference vector

$$\delta v^{\alpha} \approx R^{\alpha}{}_{\beta\mu\nu} v^{\beta} ds^{\mu\nu} . \tag{13}$$

Therefore, the parallel transport of a vector induces a rotation driven by curvature. For further reference let us also introduce here the homothetic curvature tensor, defined as

$$g^{\alpha\beta} R_{\alpha\beta\mu\nu} = R^{\alpha}{}_{\alpha\mu\nu} \equiv \Omega_{\mu\nu} , \tag{14}$$

which shall be useful later. The curvature can be decomposed into eleven irreducible components. In the language of forms, six of these belong to the antisymmetric part of the curvature 2-form $R_{[ab]}$ (36 components) while the other five constitute the symmetric part $R_{(ab)}$ (60 components), which vanishes in the absence of non-metricity.

2.2.2. Torsion

The torsion tensor can be introduced in holonomic coordinates as the antisymmetric part of the affine connection $\Gamma^{\alpha}{}_{[\beta\gamma]}$. It has 24 independent components given by

$$T^{\alpha}_{\beta\gamma} \equiv \Gamma^{\alpha}{}_{[\beta\gamma]} . \tag{15}$$

If we apply it to the vectors U and V we obtain the new vector field

$$T(U,V) = \nabla_U V - \nabla_V U - [U,V] . \tag{16}$$

This expression allows to visualize the geometrical effect of torsion. Indeed, starting from a given point on the manifold and parallel-transporting V along the (integral curve of) U, through an infinitesimal distance, and doing the complementary trip with U, that is, parallel-transporting it along V, the expected parallelogram does not close if the manifold has a non-vanishing torsion. The two end points are separated from each other by a (spacetime) translation, given by

$$\xi^{\alpha} \approx 2T^{\alpha}_{\mu\nu} ds^{\mu\nu} \,. \tag{17}$$

The torsion field has three irreducible pieces completing the 24 independent components, $T^{\lambda}_{\mu\nu} = \bar{T}^{\lambda}_{\mu\nu} + \frac{2}{3}\delta^{\lambda}_{[\nu} T_{\mu]} + g^{\lambda\sigma}\epsilon_{\mu\nu\sigma\rho}\check{T}^{\rho}$, where the traceless tensor (16 components) obeys $\bar{T}^{\lambda}_{\mu\lambda} = 0$ and $\epsilon^{\lambda\mu\nu\rho}\bar{T}_{\mu\nu\rho} = 0$, while T_{μ} is the trace vector and $\check{T}^{\lambda} \equiv \frac{1}{6}\epsilon^{\lambda\alpha\beta\gamma}T_{\alpha\beta\gamma}$ the pseudo-trace (axial) vector.

2.2.3. Non-Metricity

In holonomic coordinates the non-metricity tensor can be defined as the covariant derivative of the spacetime metric $g_{\beta\gamma}$. It has 40 independent components

$$Q_{\alpha\beta\gamma} = \nabla_{\alpha} g_{\beta\gamma} \,. \tag{18}$$

The trace-vector part $Q_{\mu} \equiv Q_{\mu\alpha}{}^{\alpha}$ is the only non-vanishing part of non-metricity in a Weyl geometry (e.g., Riemann-Weyl or Cartan-Weyl, to be discussed later), known as the Weyl co-vector. It is actually related to the homothetic curvature (14) via the equation

$$\Omega_{\mu\nu} = -\frac{1}{2}\left(\partial_{\mu} Q_{\nu} - \partial_{\nu} Q_{\mu}\right) \,. \tag{19}$$

After the parallel transport of a vector along an infinitesimal closed loop, there is a change in length given by

$$\delta l \approx l(v)\Omega_{\mu\nu} ds^{\mu\nu} \,. \tag{20}$$

Indeed, if Q_{μ} is the only non-vanishing part of $Q_{\alpha\beta\gamma}$, then $\nabla_{\alpha} g_{\beta\gamma} \sim Q_{\alpha} g_{\beta\gamma}$ and the norm of vectors change due to this Weyl co-vector as

$$\nabla_{\nu}(V^2) \sim Q_{\nu} V^2, \qquad V^2 = g_{\beta\gamma} V^{\beta} V^{\gamma} \,. \tag{21}$$

Therefore, under the presence of non-metricity the parallel transport of a vector involves a change on its length.

From the definition of the non-metricity tensor, one can derive the following Bianchi identity-type relation:

$$\nabla_{[\mu} Q_{\nu]\alpha\beta} = -R_{(\alpha\beta)\mu\nu} + Q_{\lambda\alpha\beta} T^{\lambda}{}_{\mu\nu} \,. \tag{22}$$

Therefore, if $Q_{\alpha\beta\gamma} \neq 0$ then $R_{(\alpha\beta)\mu\nu} \neq 0$. The quantity $R_{(\alpha\beta)\mu\nu}$ can be identified as the non-Riemannian part of the curvature, and its related to a non-Lorentzian linear connection and the breaking of Lorentz invariance. Finally, non-metricity can be decomposed into its trace-vector part and the traceless tensor part $\bar{Q}_{\alpha\beta\gamma}$, according to $Q_{\alpha\beta\gamma} = \bar{Q}_{\alpha\beta\gamma} + \frac{1}{4} g_{\beta\gamma} Q_{\alpha}$. In the language of forms, the latter can be further decomposed into a shear co-vector and a shear 2-form part in such a way that at the end of the day the tensor valued non-metricity 1-form has four irreducible pieces with respect to the Lorentz group.

To summarize the geometrical interpretation of the three pieces of the connection, one can say that they are associated to changes in the properties of a vector when parallel-transported: curvature yields a rotation, torsion a non-closure of its parallelograms, and non-metricity a change on its length. In turn, these geometrical interpretations have a deep impact in the physical models built upon any such pieces, as we learned decades ago from the physics of solid state systems with defects [32].

2.3. A Brief Note on the Conformal and Metric Structures of Spacetime Geometry

Important developments in geometrical methods in field theories suggest that the spacetime metric might not be considered as a fundamental field. In fact, the conformally invariant part of the metric can be derived from local and linear electrodynamics [33–37]. The pre-metric approach to electrodynamics give fully general, coordinate-free, covariant inhomogeneous equations

$$dH = J\,, \tag{23}$$

and homogeneous field equations

$$dF = 0\,, \tag{24}$$

from charge conservation and magnetic flux conservation, respectively. Since there is no metric involved, electrodynamics is not linked to a Minkowski spacetime at a fundamental level. In order to close the system the postulate on the constitutive relations $H = H(F)$ is required which, in vacuum, can be interpreted as constitutive relations for the spacetime itself [38]. On the other hand, linear, local and homogeneous constitutive relations $H = \lambda \star F$ introduce the spacetime metric[4], which is also involved in the more general linear-type relations $H_{\alpha\beta} = \kappa^{\mu\nu}_{\alpha\beta} F_{\mu\nu}$. Since the excitation H and the field strength F are 2-forms, the tensor $\kappa^{\mu\nu}_{\alpha\beta}$ has 36 independent components characterizing the electromagnetic (propagation) properties of spacetime.

From pre-metric electrodynamics plus linear, local, homogeneous constitutive relations, one can derive the conformally invariant part of the spacetime metric. Once the propagation of electromagnetic fields is considered and the geometrical optics limit is taken, one finds a quartic Fresnell surface which becomes a quadratic surface under the imposition of zero birefringence (double refraction) in vacuum. This quadratic surface defines the light-cone. Therefore, pre-metric electrodynamics, together with linear, local, homogeneous constitutive spacetime-electromagnetic relations and zero birefringence, gives the spacetime metric up to a conformal factor. The conformally invariant part of the metric (the causal structure) is derived from linear electrodynamics.

We can conclude this part by stating that the resulting light cone or causal-electromagnetic structure is a conformal geometry with local conformal symmetries associated to the light cone at each spacetime point. In such a geometry, if one parallel-transports one light cone from a given point to a neighbouring point, it will be deformed according to the non-metricity tensor, which is linked to the existence of a non-Lorentzian linear connection ($\Gamma^{ab} \neq -\Gamma^{ba}$). This alone is sufficient to break Lorentz symmetry. One sees very clearly that, in addition to the metric not being fundamental, there is a primacy of the conformal (over the metric) structure, which itself is an electromagnetic derived quantity. Moreover, electrodynamics is fundamentally connected to the conformal geometrical structure (and to the conformal group), but neither to the Poincaré/Lorentz group nor to Minkowski spacetime.

3. Metric-Affine Formalism and Classical Spacetime

3.1. A Brief Outlook on Metric-Affine Gravity

Let us now summarize the fundamental structures of spacetime and their relation to symmetry groups. We have the fundamental 1-forms, the linear co-frame θ^a and the linear connection $\Gamma^a{}_b$, which are the potentials for the 4-dimensional translations $T(4)$ and the general linear $GL(4,\Re)$ groups, respectively. The corresponding field strengths correspond to well known objects from differential geometry, namely, the torsion $T^a = D\theta^a = d\theta^a + \Gamma^a{}_b \wedge \theta^b$ and curvature $R^a{}_b = d\Gamma^a{}_b + \Gamma^a{}_c \wedge \Gamma^c{}_b$ 2-forms, respectively. The metric is introduced as the 0-form potential with the corresponding field

[4] In d dimensions the Hodge star operator maps p-forms to $(d-p)$-forms. In components we get $H_{\mu\nu} = \frac{\lambda}{2}\sqrt{-g}g^{\alpha\lambda}g^{\beta\gamma}\epsilon_{\mu\nu\alpha\beta}F_{\lambda\gamma}$.

strength being the non-metricity 1-form $Q_{ab} = Dg_{ab}$. The linear frame establishes a link between the (symmetries on the) local tangent fibers and the spacetime manifold while the linear connection can be viewed as a guidance field reflecting the inertial character for matter fields propagating on the spacetime manifold. Finally, the metric allows the determination of spatial and temporal distances and angles. The spacetime geometry with all these structures is called a *metric-affine geometry* (MAG), with non-vanishing torsion, curvature and non-metricity, and its fundamental local group of spacetime symmetries is the affine group $A(4,\Re) = T(4) \rtimes GL(4,\Re)$, which is the semi-direct product of the group of translations and the general linear group.

A truly independent linear connection is given by the decomposition in (6) that is useful to analyse the relation between non-Lorentzian metrics, the breaking of Lorentz invariance and the presence of non-metricity. Since $Q_{ab} = 2N_{(ab)}$, if non-metricity is zero, then the connection is Lorentzian and the spacetime geometry is the RC one with curvature and torsion. Such a spacetime is fundamentally linked to the local symmetry group of Poincaré transformations $P(1,3) = T(4) \rtimes SO(1,3)$, which is the semi-direct product between the translations $T(4)$ and the Lorentz group $SO(1,3)$. As one can see from the expressions for the field strengths (torsion and curvature), the connection also enters in the expression for the field strength of the co-frame. This term is unavoidably present and is due to the semi-direct product[5] structure of the Poincaré group (or the affine group) and, therefore, curvature and torsion are somehow intertwined.

In the self-consistent metric-affine formalism, the gravitational interaction is described as a gauge theory of the affine group $A(4,\Re)$, together with the assumption of a metric, with the potentials (θ^a, Γ^{ab}) coupled to the corresponding Noether currents $(\tau^a, \Delta^a{}_b)$. The latter are the vector-valued canonical energy-momentum $\tau^a = \delta\mathcal{L}_{mat}/\delta\theta_a$ and the tensor-valued hypermomentum $\Delta^a_b = \delta\mathcal{L}_{mat}/\delta\Gamma^b_a$ 3-form currents[6], respectively, while the metric g_{ab} couples to the symmetric (Hilbert) energy momentum $T_{ab} = 2\delta\mathcal{L}_{mat}/\delta g^{ab}$. The hypermomentum can be decomposed according to the expression

$$\Delta_{ab} = s_{ab} + \frac{1}{4}g_{ab}\Delta^c{}_c + \bar{\Delta}_{ab}\,, \tag{25}$$

including the spin $s_{ab} = -s_{ba}$, the dilatation $\Delta^c{}_c$, and the shear $\bar{\Delta}_{ab}$ currents. The Noether currents are the fundamental sources of MAG. A truly independent connection in MAG can be written as

$$\Gamma^{ab} = \Gamma^{[ab]} + \frac{1}{4}g^{ab}\Gamma^c{}_c + \left(\Gamma^{ab} - \frac{1}{4}g^{ab}\Gamma^c{}_c\right), \tag{26}$$

including the Lorentzian piece $\Gamma^{[ab]}$, the trace part $\frac{1}{4}g^{ab}\Gamma^c{}_c$ and the shear part $\Gamma^{ab} - \frac{1}{4}g^{ab}\Gamma^c{}_c$. The Lorentzian connection couples to the spin current and the trace and shear parts couple to the dilatation and shear currents, respectively. As previously said it is the non-Lorentzian part of the connection that imply the non-vanishing of non-metricity, therefore, the dilatation and shear hypermomentum currents are intimately related to the non-metricity of metric affine spacetime geometry.

The variational principle is applied to the action of this theory (including the gravitational part and the matter Lagrangian) by varying it with respect to the gauge potentials of the affine group, (θ^a, Γ^{ab}). This leads to two sets of dynamical equations, while a third set of equations is obtained by varying the action with respect to the metric potential g_{ab}. At the end, the dynamics is described only via two sets of equations, since the gravitational equation obtained by varying with respect to

---

5 The semi direct product implies that the generators of $T(4)$ and $GL(4,\Re)$ (or $SO(1,3)$) do not commute.

6 These currents can be represented as 1-forms or as 3-forms. In fact, in the gauge approach to gravity they emerge naturally from Noether equations as 3-forms, being natural objects for integration over volumes. As 1-forms one can write $\tau^a = \tau^a_\mu dx^\mu$ and $\Delta^a{}_b = \Delta^a{}_{b\mu}dx^\mu$. This map between 3-form or 1-form representation is related to the fact that in d-dimension a k-form has $\frac{d!}{k!(d-k)!}$ independent components. In four dimensions both 3-forms and 1-forms have four independent components.

the metric potential or the one obtained by variation with respect to the translational potential can be dropped out, as long as the other gravitational equation (derived from variation with respect to the linear connection) is fulfilled. This procedure and the fundamental quantities and relations here exposed summarizes the basics of the MAG formalism.

3.2. *Classical Spacetime Paradigms*

We now revise some important classical spacetime paradigms, which play a fundamental role in the formulation of gravitational theories.

3.2.1. Minkowski Spacetime—M_4

This is a pseudo-Euclidean geometry with vanishing curvature, torsion and non-metricity. It has global/rigid Poincaré symmetries and a globally absolute causal structure. For inertial reference systems the connection vanishes. The inertial frame and the inertial properties of matter are defined with respect to this absolute spacetime. In particular, the mass and spin of particles can be considered to be intrinsic to particles and are classified with the help of the Casimir operators using the irreducible representations of the Poincaré group.

3.2.2. (Pseudo)Riemann Geometry of GR—V_4

A post-Euclidean geometry with curvature and vanishing torsion and non-metricity. It obeys local Lorentz symmetries and in local geodesic frames the (Levi-Civita) connection vanishes but (the Weyl part of) curvature is non-zero. The causal structure $ds^2 = 0$ is locally invariant under local Lorentz symmetries. The inertial properties of matter are locally defined with respect to absolute spacetime.

3.2.3. Riemann-Cartan Geometry—U_4

This geometry has curvature and torsion but zero non-metricity, therefore the connection is Lorentzian (spin connection). It obeys local Poincaré $P(1,3)$ symmetries and the causal structure is locally invariant under such a group. The inertial properties of matter are locally defined with respect to absolute spacetime. The corresponding curvature and its contractions (Ricci tensor and Ricci scalar in the RC geometry) are given by the expressions

$$\begin{aligned} R^{\alpha}{}_{\beta\mu\nu} &= \tilde{R}^{\alpha}{}_{\beta\mu\nu} + \tilde{\nabla}_{\mu} K^{\alpha}{}_{\beta\nu} - \tilde{\nabla}_{\nu} K^{\alpha}{}_{\beta\mu} + K^{\alpha}{}_{\lambda\mu} K^{\lambda}{}_{\beta\nu} - K^{\alpha}{}_{\lambda\nu} K^{\lambda}{}_{\beta\mu} \\ R_{\beta\nu} &= \tilde{R}_{\beta\nu} + \tilde{\nabla}_{\alpha} K^{\alpha}{}_{\beta\nu} - \tilde{\nabla}_{\nu} K^{\alpha}{}_{\beta\alpha} + K^{\alpha}{}_{\lambda\alpha} K^{\lambda}{}_{\beta\nu} - K^{\alpha}{}_{\lambda\nu} K^{\lambda}{}_{\beta\alpha}, \\ R &= \tilde{R} - 2\tilde{\nabla}^{\lambda} K^{\alpha}{}_{\lambda\alpha} + g^{\beta\nu}\left(K^{\alpha}{}_{\lambda\alpha} K^{\lambda}{}_{\beta\nu} - K^{\alpha}{}_{\lambda\nu} K^{\lambda}{}_{\beta\alpha}\right), \end{aligned} \tag{27}$$

respectively. The curvature tensor obeys the following first and second Bianchi identities

$$\nabla_{[\gamma} R^{\alpha}{}_{\beta|\mu\nu]} = 2R^{\alpha}{}_{\beta\lambda[\mu} T^{\lambda}{}_{\nu\gamma]}, \tag{28}$$

$$R^{\alpha}{}_{[\beta\mu\nu]} = -2\nabla_{[\nu} T^{\alpha}{}_{\beta\mu]} + 4T^{\alpha}{}_{\lambda[\beta} T^{\lambda}{}_{\mu\nu]}. \tag{29}$$

These relations can be deduced from the corresponding expressions in terms of the curvature and torsion 2-forms, namely[7]

$$DR^{a}{}_{b} = 0, \qquad DT^{c} = R^{c}{}_{d} \wedge \theta^{d}. \tag{30}$$

The anholonomic or frame indices are also called symmetry indices since they are related to the tangent fibers where the local spacetime symmetries are characterized. In this case these are known also as Lorentz indices. Spin connections are related to rotations (two Lorentz indices) and the tetrads

[7] Here $DR^{a}{}_{b} = dR^{a}{}_{b} + \Gamma^{a}{}_{c} \wedge R^{c}{}_{b} + \Gamma^{c}{}_{b} \wedge R^{a}{}_{c}$ and $DT^{c} = dT^{c} + \Gamma^{c}{}_{d} \wedge T^{d}$.

or co-frames are related to translations (one Lorentz index). The same is valid for the corresponding field strengths. The symmetry indices have a direct link to geometrical interpretations via the strong relation between group theory and geometry. Two symmetry indices for curvature means that it is related to rotations, and one symmetry index for torsion means that it is related to translations. On the other hand, since both curvature and torsion are represented by 2-forms, as geometrical objects these are therefore connected to 2-surfaces. One can thus say that Cartan pictured a RC geometry by associating to each infinitesimal surface element a rotation and a translation.

The components of the curvature and torsion 2-forms in terms of the tetrads and spin connection are

$$R^a{}_{b\mu\nu} = \partial_\mu \Gamma^a{}_{b\nu} - \partial_\nu \Gamma^a{}_{b\mu} + \Gamma^a{}_{c\mu}\Gamma^c{}_{b\nu} - \Gamma^a{}_{d\nu}\Gamma^d{}_{b\mu}\,, \tag{31}$$

$$T^a{}_{\mu\nu} = \partial_\mu \theta^a{}_\nu - \partial_\mu \theta^a{}_\nu + \Gamma^a{}_{b\mu}\theta^b{}_\nu - \Gamma^a{}_{b\nu}\theta^b{}_\mu\,, \tag{32}$$

respectively. On the other hand, the components of the 1-form spin connection

$$\Gamma^a{}_{b\nu} = \tilde{\Gamma}^a{}_{b\nu} + K^a{}_{b\nu}\,, \tag{33}$$

where $K^a{}_{b\mu}$ are the components of the contortion 1-form, are related to the holonomic (spacetime) components of the affine connection through the relations

$$\Gamma^a{}_{b\nu} = \theta^a{}_\mu \partial_\nu e_b^{\ \mu} + \theta^a{}_\mu \Gamma^\mu{}_{\beta\nu} e_b^{\ \beta} \tag{34}$$

$$\Gamma^\lambda{}_{\nu\mu} = e_a^{\ \lambda}\partial_\mu \theta^a{}_\nu + e_a^{\ \lambda}\Gamma^a{}_{b\mu}\theta^b{}_\nu\,. \tag{35}$$

Since the connection characterizes the way the linear frame/co-frame changes from point to point, these relations can be deduced from the equation

$$\partial_\nu e_b^{\ \mu} + \Gamma^\mu{}_{\beta\nu} e_b^{\ \beta} \equiv \Gamma^c{}_{b\nu} e_c^{\ \mu}\,, \tag{36}$$

which expresses the fact that the total covariant derivative of the tetrads with respect to both holonomic and anholonomic (Lorentz) indices is vanishing, that is, $\partial_\nu e_b^{\ \mu} + \Gamma^\mu{}_{\beta\nu} e_b^{\ \beta} - \Gamma^c{}_{b\nu} e_c^{\ \mu} = 0$ and $\partial_\mu \theta^a{}_\nu - \Gamma^\beta{}_{\nu\mu}\theta^a{}_\beta + \Gamma^a{}_{b\mu}\theta^b{}_\nu = 0$. By changing from one spacetime point to another in the frame/coframe, the tetrads also change. The new tetrads $\bar{e}_b^{\ \mu}$ can be expressed in terms of the original ones as $\bar{e}_b^{\ \mu} = \Lambda^a_b e_a^{\ \mu}$, and by using the relations

$$\eta_{ab} = e_a^{\ \mu} e_b^{\ \nu} g_{\mu\nu}, \quad g_{\mu\nu} = \theta^a{}_\mu \theta^b{}_\nu \eta_{ab}, \quad \sqrt{-g} = det(\theta^a{}_\mu), \tag{37}$$

one can show that Λ^a_b are Lorentz matrices, that is, the tetrads undergoes a Lorentz rotation under the motion from one spacetime point to another. For that reason, the linear connection, which characterizes the change in the frame/co-frame, is called Lorentzian or spin connection. As previously said, the six Lorentzian connections are the potentials associated to the generators of the Lorentz group.

The matrices $e_a^{\ \mu}$ and their inverses, which establish a correspondence between the local spacetime metric and the Minkowski metric on the tangent/cotangent planes, constitute a map between holonomic and anholonomic bases. For example, for vectors

$$v^\mu = v^a e_a^{\ \mu}, \quad v^b = v^\nu \theta^b{}_\nu, \quad h_\mu = h_a \theta^a{}_\mu, \quad h_b = h_\nu e_b^{\ \nu}. \tag{38}$$

Since the tangent and co-tangent spaces, $T_p(\mathcal{M})$ and $T_p^*(\mathcal{M})$, change while moving from one point on the manifold to another, the notion of covariant derivative is extended for quantities with Lorentz indices. This is done with the spin connection. For (Lorentz) vectors and co-vectors we have

$${}^s\nabla_\mu v^a = \partial_\mu v^a + \Gamma^a{}_{b\mu} v^b, \quad {}^s\nabla_\nu h_c = \partial_\nu h_c - \Gamma^d{}_{c\nu} h_d. \tag{39}$$

The anholonomic basis e_a and θ^b also characterize the local spacetime symmetries in the tangent fibers, since the local symmetry group algebra is implicit in the relations

$$[e_a, e_b] = f^c_{\ ab} e_c, \qquad f^c_{\ ab} = e_a^{\ \mu} e_b^{\ \nu} (\partial_\nu \theta^c_{\ \mu} - \partial_\mu \theta^c_{\ \nu}), \tag{40}$$

where the (group) structure constants $f^a_{\ bc}$ are known as Ricci rotation coefficients.

In RC geometry the affine (self-parallel) geodesics differ, in general, from the extremal geodesics and are given by

$$\frac{du^\alpha}{ds} + \Gamma^\alpha_{\ \beta\gamma} u^\beta u^\gamma = 0 \quad \Leftrightarrow \quad \frac{du^\alpha}{ds} + \tilde{\Gamma}^\alpha_{\ \beta\gamma} u^\beta u^\gamma = -K^\alpha_{\ (\beta\gamma)} u^\beta u^\gamma, \tag{41}$$

respectively, where u^α are the components of the (4-velocity) tangent vector to the curve. However, in the RC spacetime matter particles with vanishing intrinsic spin are insensitive to the non-Riemannian part of the geometry and, therefore, to torsion. Such particles follow the extremal paths computed from the Levi-Civita connection. Moreover if torsion is completely antisymmetric, as in the case of ECSK theory, then $K^\alpha_{\ (\beta\gamma)} = 0$ and the extremal and self-parallel geodesics coincide. In any case, the appropriate evaluation of the motion of particles with intrinsic spin (fermions) in a RC spacetime should be performed from an analysis of the corresponding Dirac equation and then proceeding with a classical approximation, for instance, using a WKB method.

As we mentioned before, since curvature and torsion are 2-forms, one can imagine the RC geometry as having at each point an associated infinitesimal surface element with a rotational (curvature) and translational (torsion) transformation. The picture of a RC manifold with a discrete structure can naturally emerge from imposing a finite minimum surface element (rater than infinitesimal elements), which would imply, to some extent, that torsion and curvature would become quantized. Indeed, as we shall see, the mathematical methods of Cartan's exterior calculus (differential forms) within gauge theories of gravity contribute to clarify the appropriate physical degrees of freedom and mathematical objects that should be quantized in a quantum (Yang-Mills) gauge theory of gravity. This procedure, in this formalism of forms, can be done in a metric-free (pre-metric) way, avoiding the difficulties often found in perturbative approaches that require some (well behaved) background (spacetime-vacuum), with respect to which the perturbations can be defined. Since any such gravity theory should be (metric) background independent, this is usually a huge problem, since the background is also the very thing one would like to quantize and should come as a solution of the dynamical equations. Gravity in exterior forms can shed some light on this challenge, at least by establishing a metric-independent framework and a well identified set of canonically conjugate variables to be quantized. As we shall see later, in Yang-Mills type of gauge theories of gravity this will not require the full metric structure, but only the conformally invariant part of the metric, that is the conformal-causal structure. The latter is introduced via the Hodge star operator in the constitutive relations between the field strengths and the conjugate momenta.

3.2.4. Riemann-Weyl Geometry—W_4

This is the spacetime geometry implicit in Weyl's gauge theory for unifying gravity and electromagnetism by extending the local Lorentz symmetries to include dilatations. It has curvature and (the trace-vector part of) non-metricity but vanishing torsion, and it obeys local symmetries under the Weyl group $W(1,3)$, which includes the $P(1,3)$ group and dilatations. The causal structure is locally invariant under the $W(1,3)$ group, but the spacetime (metric) is not absolute, changing under dilatation-type of coordinate transformations. Accordingly, the inertial properties of matter cannot be defined with respect to an absolute metric structure. One may postulate that matter is endowed with conformally-invariant physical properties.

The Weyl connection is given by $\Gamma^{\alpha}{}_{\beta\nu} = \tilde{\Gamma}^{\alpha}{}_{\beta\nu} + q^{\alpha}{}_{\beta\nu}$, where the distortion tensor in this case is related to the Weyl co-vector $Q_{\mu} \equiv Q_{\mu\lambda}{}^{\lambda}$ as

$$q^{\alpha}{}_{\beta\nu} \equiv \frac{1}{2}(\delta^{\alpha}_{\beta} Q_{\gamma} + \delta^{\alpha}_{\gamma} Q_{\beta} - Q^{\alpha} g_{\beta\gamma}) \,, \tag{42}$$

while the curvature of the Weyl connection is given by

$$R^{\alpha}{}_{\beta\mu\nu} = \tilde{R}^{\alpha}{}_{\beta\mu\nu} + \tilde{\nabla}_{\mu} q^{\alpha}{}_{\beta\nu} - \tilde{\nabla}_{\nu} q^{\alpha}{}_{\beta\mu} + q^{\alpha}{}_{\lambda\mu} q^{\lambda}{}_{\beta\nu} - q^{\alpha}{}_{\lambda\nu} q^{\lambda}{}_{\beta\mu}. \tag{43}$$

The Weyl co-vector is the trace vector part of the non-metricity and, therefore, there are scale (dilatations) type of distortions in the geometry, implicit in the relation $\nabla_{\alpha} g_{\beta\gamma} \sim Q_{\alpha} g_{\beta\gamma}$ and, as previously mentioned, the length of vectors change in this geometry as $\nabla_{\nu}(V^2) \sim Q_{\nu} V^2$. Weyl identified the Weyl co-vector as the electromagnetic 4-potential, and the previously defined homothetic curvature (14) as the electromagnetic Faraday tensor. We stress that the pre-metric foundations of electrodynamics indeed show the deep connection between electromagnetism and conformal geometry and conformal symmetries, which presuppose a natural framework for the breaking of Lorentz symmetry.

3.2.5. Riemann-Cartan-Weyl Geometry—Y_4

This is a generalization of the Weyl geometry by including torsion, which is therefore the appropriate manifold for the Weyl gauge theories of gravity. Only matter with hypermomentum (25), that includes the spin current and the dilatations current, can be sensitive to torsion and to the (Weyl-covector part of) non-metricity.

3.2.6. Metric-Affine Geometry (L_4, g)

This is the richest kind of geometry considered here, having non-vanishing curvature, torsion and non-metricity. The natural symmetry group is the affine group $A(4, \Re)$. The affine connection can be written as $\Gamma^{\alpha}{}_{\beta\nu} = \tilde{\Gamma}^{\alpha}{}_{\beta\nu} + N^{\alpha}{}_{\beta\nu}$, where the distortion tensor $N^{\alpha}{}_{\beta\nu} = K^{\alpha}{}_{\beta\nu} + L^{\alpha}{}_{\beta\nu}$ includes both contortion and disformation, as described in Section 2.2. The curvature can be written as

$$R^{\alpha}{}_{\beta\mu\nu} = \tilde{R}^{\alpha}{}_{\beta\mu\nu} + \tilde{\nabla}_{\mu} N^{\alpha}{}_{\beta\nu} - \tilde{\nabla}_{\nu} N^{\alpha}{}_{\beta\mu} + N^{\alpha}{}_{\lambda\mu} N^{\lambda}{}_{\beta\nu} - N^{\alpha}{}_{\lambda\nu} N^{\lambda}{}_{\beta\mu}, \tag{44}$$

or, alternatively, as

$$R^{\alpha}{}_{\beta\mu\nu} = \bar{R}^{\alpha}{}_{\beta\mu\nu} + \bar{\nabla}_{\mu} L^{\alpha}{}_{\beta\nu} - \bar{\nabla}_{\nu} L^{\alpha}{}_{\beta\mu} + L^{\alpha}{}_{\lambda\mu} L^{\lambda}{}_{\beta\nu} - L^{\alpha}{}_{\lambda\nu} L^{\lambda}{}_{\beta\mu}, \tag{45}$$

where $\bar{R}^{\alpha}{}_{\beta\mu\nu}$ and $\bar{\nabla}_{\mu}$ are the curvature and covariant derivative of a Riemann-Cartan connection.

The above discussion represents just a quick excursion in the *affinesia* world[8], where different geometries emerge depending on the pieces of the connection one decides to keep/remove. In the next section, we shall draw our attention to the gauge formulation of gravity and its relation to all these spacetime geometries.

4. Gauge Theories of Gravity

The gauge approach to gravity broadens our study of the deep relation between symmetry principles (group theory) and geometrical methods. Of special relevance for our analysis is the PGTG class, which constitutes a promising candidate for an appropriate description of classical gravity

---

8 This anecdotal notion was first coined by José Beltrán-Jiménez in one of his talks about the trinity of gravity [39].

including post-Einstein strong-field predictions. In order to present the structure of the gauge approach to gravity, it is useful first to revisit the Weyl-Yang-Mills formalism for gauge fields.

4.1. The Weyl-Yang-Mills Formalism

The gauge approach to Yang-Mills fields follows from two major steps, the rigid (global) symmetries of a physical system described by a matter Lagrangian, and the localization (gauging) of those symmetries.

In the first step one considers rigid (global) symmetries as follows:

- Start with a field theory: $\mathcal{L}_m = \mathcal{L}_m(\Psi, d\Psi)$ of some matter fields Ψ.
- The matter Lagrangian is invariant under some internal symmetry group, described by a (semi-simple) Lie group with generators T_a.
- Noether's first theorem implies a conserved current: $dJ = 0$.

In the second step the localization (gauging) of the symmetries is performed according to the following procedure:

- The symmetries are described on each spacetime point introducing the compensating (gauge) field $A = A^a_\mu T_a dx^\mu$.
- This is a new field that couples minimally to matter and represents a new interaction.
- To preserve the symmetries this gauge potential A transforms in a suitable way allowing to construct a (gauge) covariant derivation $d\Psi \longrightarrow D\Psi = (d + A)\Psi$.
- The Lagrangian includes this minimal coupling between the matter fields and the gauge potential $\mathcal{L}(\Psi, d\Psi) \longrightarrow \mathcal{L}(\Psi, D\Psi)$.
- The gauge potential acts on the components of the matter fields defined with respect to some reference frame. Geometrically, it is the connection of the frame bundle (fiber bundle) related to the symmetry group.
- The conservation equation is generalized as $dJ = 0 \rightarrow DJ = 0$.

In order for A to represent a true dynamical variable with its own degrees of freedom, the Lagrangian of the theory has to include a kinetic term, representing the new interaction $\mathcal{L} = \mathcal{L}_m + \mathcal{L}_A$. The invariance of $\mathcal{L}_A$ is secured by constructing it with the gauge invariant field strength $F = DA = dA + A \wedge A$ which, geometrically, can be interpreted as the curvature 2-form of the fiber bundle. Note that to have second order (on A) inhomogeneous Yang-Mills field equations, one must choose $\mathcal{L}_A = \mathcal{L}_A(F)$.

Written in terms of exterior forms, the inhomogeneous Yang-Mills equations for the gauge potential are

$$DH = J\,, \tag{46}$$

where $H = \partial\mathcal{L}/\partial F$ is the excitation 2-form, and $J = \partial\mathcal{L}_m/\partial A$ is the conserved Noether current, which acts as a source for the potential. Here, $DH \equiv dH + A \wedge H$. The homogeneous field equation corresponds to a Bianchi identity, obtained from the derivation of the potential twice, namely

$$DF = 0 \quad \Leftrightarrow \quad dF = -A \wedge F\,. \tag{47}$$

Note also that the equation for the conservation of the Noether current is generalized via the gauge covariant exterior derivative

$$DJ = 0 \quad \Leftrightarrow \quad dJ = -A \wedge J\,. \tag{48}$$

For non-abelian groups the gauge field contributes with an associated ("isospin") current, $-A \wedge H$. In such a case $dJ \neq 0$, and the (gauge) interaction field is charged, unlike the case of abelian groups, such as the $U(1)$ group of electromagnetism.

In order to have a wave-like inhomogeneous Yang-Mills (quasi-linear) equation, and paralleling the case of electromagnetism, $\mathcal{L}$ can depend quadratically with F at most and, therefore, H must depend linearly, for instance as $H = H(F) = \alpha \star F$. The Yang-Mills inhomogeneous equations then turn into

$$D \star F = d \star F + A \wedge \star F = \alpha^{-1} J \Leftrightarrow d \star F = \alpha^{-1}(J + J^A), \tag{49}$$

where $J^A \equiv -\alpha A \wedge \star F$. In this formalism one can see the clear analogies between classical mechanics and Yang-Mills field theory, which we summarize in Table 1. In particular, it is clear that the field strength F is the generalized velocity while the excitation H is the conjugate momentum. The constitutive relation $H(F)$ is implicit in the Lagrangian formulation and corresponds in perfect analogy to the functional relation between generalized velocities and conjugate momenta of classical mechanics. As an extension of these analogies, one is naturally led to identify the appropriate pairs of canonically conjugate variables that should be quantized in the corresponding (canonical) quantum field theory (see Table 2).

Table 1. The analogies between classical mechanics, Yang-Mills field and gravity-Yang-Mills theory in the language of exterior forms.

	Gravity Yang-Mills	**Yang-Mills** $\mathcal{L} = \mathcal{L}(A, DA)$	**Classical Mechanics** $\mathcal{L} = \mathcal{L}(q, \dot{q})$
Configuration variables	$(\Gamma^a_b, \vartheta^a)$	A	q
Generalized velocities	$R^a_b = D\Gamma^a_b$ $T^a = D\vartheta^a$	$F \equiv DA$	$\dot{q}$
Lagrange equations	$D\left(\frac{\partial \mathcal{L}}{\partial R^{ab}}\right) - \varsigma_{ab} = s_{ab}$ $D\left(\frac{\partial \mathcal{L}}{\partial T^a}\right) - \pi_a = \tau_a$	$D\left(\frac{\partial \mathcal{L}}{\partial F}\right) = J$ $J = \frac{\partial \mathcal{L}_m}{\partial A}$	$\frac{d}{dt}\left(\frac{\partial \mathcal{L}}{\partial \dot{q}}\right) = \frac{\partial \mathcal{L}}{\partial q}$
Conjugate momenta	$H_{ab} = \left(\frac{\partial \mathcal{L}}{\partial R^{ab}}\right)$ $H_a = \left(\frac{\partial \mathcal{L}}{\partial T^a}\right)$	$H = \frac{\partial \mathcal{L}}{\partial DA} = \frac{\partial \mathcal{L}}{\partial F}$	$p = \frac{\partial \mathcal{L}}{\partial \dot{q}}$
Constitutive relations	$H^{ab} = H^{ab}(R^{ab}, T^a)$ $\quad H^a = H^a(R^{ab}, T^a)$ $H^{ab} \sim \star R^{ab}$ $\quad H^a \sim \star T^a$ (linear)	$H = H(F)$ $H \sim \star F$ (linear)	$p = p(\dot{q})$
Canonical variables	(Γ^a_b, H^a_b) $\quad (\vartheta^a, H^a)$	(A, H)	(q, p)
Hamiltonian	$\mathcal{H} \equiv R^{ab} \wedge H_{ab} + T^a \wedge H_a - \mathcal{L}(\Gamma, \vartheta, R, T)$ $R^{ab} = R^{ab}(H^{ab}, H^a)$ $\quad T^a = T^a(H^{ab}, H^a)$	$\mathcal{H} \equiv F \wedge H - \mathcal{L}$ $F = F(H)$	$\mathcal{H} \equiv \dot{q}p - \mathcal{L}(q, \dot{q})$ $\dot{q} = \dot{q}(p)$
Hamilton equations	$D\Gamma^{ab} = \frac{\partial \mathcal{H}}{\partial H_{ab}}$ $\quad DH_{ab} = -\frac{\partial \mathcal{H}^{eff}}{\partial \Gamma^{ab}}$ $D\vartheta^a = \frac{\partial \mathcal{H}}{\partial H_a}$ $\quad DH_a = -\frac{\partial \mathcal{H}^{eff}}{\partial \vartheta^a}$	$DA = F = \frac{\partial \mathcal{H}}{\partial H}$ $DH = -\frac{\partial \mathcal{H}^m}{\partial A} = J$	$\frac{d}{dt}q = \frac{\partial \mathcal{H}}{\partial p}$ $\quad \frac{d}{dt}p = -\frac{\partial \mathcal{H}}{\partial q}$

Table 2. The analogies between canonical quantization in quantum mechanics and in the exterior calculus approach to Yang-Mills theories and gauge theories of gravity (*à la* Yang-Mills).

	Gravity Yang-Mills $\hat{\mathcal{H}} = \hat{\mathcal{H}}(\hat{\Gamma}^{ab}, \hat{\vartheta}^a, \hat{H}^{ab}, \hat{H}^a)$		**Yang-Mills** $\hat{\mathcal{H}} = \hat{\mathcal{H}}(\hat{A}, \hat{H})$	**Quantum Mechanics** $\hat{\mathcal{H}} = \hat{\mathcal{H}}(\hat{q}, \hat{p})$
Quantum operators	$\Gamma^{ab} \to \hat{\Gamma}^{ab}$ $H^{ab} \to \hat{H}^{ab}$	$\vartheta^a \to \hat{\vartheta}^a$ $H^a \to \hat{H}^a$	$A \to \hat{A}$ $H \to \hat{H} \sim -i\frac{\partial}{\partial A}$	$q \to \hat{q}$ $p \to \hat{p} = -i\hbar\frac{d}{dq}$
Commutation relations	$[\hat{\Gamma}^{ab}, \hat{H}^{ab}] \neq 0$	$[\hat{\vartheta}^a, \hat{H}^a] \neq 0$	$[\hat{A}, \hat{H}] \neq 0$	$[\hat{q}, \hat{p}] = -i\hbar$

4.2. The Gauge Approach to Gravity

The question that arises now is whether we can apply the same procedure to gravity. The approach of Yang, Mills and Utiyama went beyond the first ideas on gauge invariance introduced by Weyl. In fact, while Yang and Mills [11] extended Weyl's gauge principle to the $SU(2)$ isospin rotations in an attempt to describe nuclear interactions, Utiyama [12] extended the gauge principle to all semi-simple

Lie groups including the Lorentz group and tried to derive GR from the gauging of the Lorentz group. Although there is some validity in his approach and an undoubtedly importance of the Lorentz group in GR, his derivation is not fully self-consistent to the formal structure of a gauge field theory. This is mainly because the Noether conserved current of the Lorentz group is not the energy-momentum, which Utiyama forced to be the source of gravity in order to obtain GR.[9] Although these efforts did not include gravity consistently, it revealed that, on a fundamental level, gauge symmetries lie at the heart of modern field theories of physical interactions. Nevertheless, this gauge formalism eventually returned to gravity when in the 60's Sciama and Kibble localized the Poincaré group of spacetime symmetries and in this way managed to show that gravity can also be consistently described as a gauge theory. Indeed, the analogies with gauge Yang-Mills theories can be easily established, as summarized in Table 3.

Table 3. The analogies between Yang-Mills fields and gravity-Yang-Mills theory in the language of exterior forms.

	Gravity Yang-Mills $\mathcal{L} = \mathcal{L}_g + \mathcal{L}_m$ $\mathcal{L}_g = \mathcal{L}_g(g_{ab}, \vartheta^a, D\Gamma^a_b, D\vartheta^a)$	**Yang-Mills** $\mathcal{L} = \mathcal{L}_A + \mathcal{L}_m$ $\mathcal{L} = \mathcal{L}(A, DA)$
gauge potentials	$(\Gamma^a_b, \vartheta^a)$ 1-forms	A 1-form
Field strengths	$R^a_b = D\Gamma^a_b$ $T^a = D\vartheta^a$	$F \equiv DA$
Symmetry group	$PGTG : SO(1,3) \rtimes$ $T(4)$ $MAG : GL(4, \Re) \rtimes$ $T(4)$	$SU(N)$
Noether currents (sources)	$PGTG$: s^{ab}_μ canonical spin density, $s_{ab} \equiv \delta\mathcal{L}_m/\delta\Gamma^{ab}$ τ^a_μ canonical energy-momentum density, $\tau_a \equiv \delta\mathcal{L}_m/\delta\vartheta^a$ MAG : Δ^{ab}_μ Hypermomentum, $\Delta_{ab} \equiv \delta\mathcal{L}_m/\delta\Gamma^{ab}$ τ^a_μ canonical energy-momentum density, $\tau_a \equiv \delta\mathcal{L}_m/\delta\vartheta^a$	$J = \frac{\partial \mathcal{L}_m}{\partial A}$ (elec. charge, isospin,...)
Excitations	$PGTG$: $H_{ab} = -\delta\mathcal{L}_A/\delta R^{ab}$ $H_a = \delta\mathcal{L}_A/\delta T^a$	$H = -\delta\mathcal{L}_A/\delta F$
Field equations	$PGTG$: $DH_{ab} - \varsigma_{ab} = s_{ab}$ $DH_a - \pi_a = \tau_a$	$DH = J$
Bianchi identities	$PGTG$: $dR^a_b + \Gamma^a_c \wedge R^c_b = -R^a_c \wedge \Gamma^c_b \quad (DR^a_b = 0)$ $DT^a = R^a_c \wedge \vartheta^c$	$dF = -F \wedge A$ $(DF = 0)$

One of the most remarkable features of the gauge approach to gravity is the intimate link between group considerations and spacetime geometry. Non-rigid (local) spacetime symmetries require non-rigid (non-Euclidian) geometries. Moreover, as previously mentioned, by extending the symmetry

[9] From a gauge theoretical perspective, the Lorentz group is not the symmetry group of Einstein's gravity. It became clear some years later that the apropriate way to derive GR from a gauge principle is to consider it as a translational gauge theory of gravity. This class of theories lives on a Weitzenböck spacetime geometry with torsion and vanishing curvature and non-metricity. These geometries and its use in attempts for a unified classical field theory were worked out by Weitzenböck, Cartan and Einstein, for example, during the first period of the so-called teleparallel formulation gravity (up to 1938). A second period in the 60's by Moller and others rekindled the interest in such theories which have more recently re-gained much attention, particularly via its $f(T)$ extensions (see e.g., Reference [40]).

group one is led to extend the spacetime geometry as well and, in this way, post-Riemann geometries have a natural place within gauge theories of gravity. For instance, while the translational gauge theories (TGTG) include non-vanishing torsion but zero curvature and non-metricity, Poincaré gauge gravity requires a RC geometry, and both Weyl(-Cartan) gauge gravity (WGTG) and conformal gauge gravity (CGTG) live on subsets of the more general metric-affine geometry with curvature, torsion and non-metricity. In particular, in WGTG the traceless part of the non-metricity vanishes. Both the 10-parametric Poincaré group and the 11-parametric Weyl group are non-simple, meaning that they can be divided into two smaller groups (a non-trivial normal sub-group and the corresponding quotient group) and the natural extension from the corresponding theories of gravity into one with a simple group leads to CGTG. The 15-parametric conformal group $C(1,3)$ is simple (its only normal sub-groups are the trivial group and the group itself), but this extension requires a generalization of Kibble's gauge procedure, due to the fact that although locally $C(1,3)$ is isomorphic to $SO(2,4)$ its realization in M_4 (Minkowski spacetime) is non-linear [41].

The PGTG can further be extended into the de Sitter or anti-de Sitter (A)dS gauge theories of gravity by localizing the $SO(1,4)$ or $SO(2,3)$ groups, respectively. Due to the fact that the (A)dS space is a maximally symmetric space which can be embedded into 5-dimensional Minkowski space (with two or one time coordinates for AdS or dS, respectively), its isometries obey Lorentz type of algebra. Under a specific limit (by setting $l \to \infty$, where l is a parameter of the group algebra), the group goes into the Poincaré algebra. Depending on the choices of the Lagrangian, one can then have explicit [42] or spontaneous [43] symmetry breaking from $SO(2,3)$ to $SO(1,3)$, for instance.

Another important class of extensions requires going beyond the Lie algebra by considering the algebra with anticommutators, in order to arrive at the super-Poincaré group [44] containing the usual Poincaré generators and proper supersymmetry (SUSY) transformations. These are generated by a Majorana spinor which acts as the (anticommuting) generator of the transformations between fermions and bosons. The simple (with one supersymmetry generator) AdS supersymmetry generalizes the simple super-Poincaré algebra although it has the same generators, and it goes back to the super-Poincaré algebra under the same limit as the AdS group goes back into the PGTG. Further extensions include the consideration of a number $1 < N < 8$ of supersymmetry generators. The gauging of these super-algebras lead directly to the bosonic gravity sector and therefore, supergravity (SUGRA) is an important class of supersymmetric gauge theories of gravity, extremely relevant for unification methods of bosons and fermion by the link it establishes between external (spacetime) symmetries and internal symmetries. In the self-consistent gauge approach, this class of theories needs to take into account post-Riemannian spacetime geometries, although many of the approaches have been done within the Riemannian geometry [24].

To illustrate the structure of the gauge approach to gravity we next consider the PGTG in more detail.

4.3. The Gravity Yang-Mills Equations of Poincaré Gauge Theories of Gravity

By applying to gravity a similar procedure as that of the Yang-Mills approach to gauge fields, one arrives at the mathematical structure of gauge theories of gravity. One starts with the (rigid) global symmetries of a matter Lagrangian with respect to a specific group of spacetime coordinate transformations, and the conserved Noether currents are identified. Then, by localizing (gauging) the symmetry group, the gauge gravitational potentials are introduced as well as the gauge covariant derivative and the respective field strengths, which are well known objects from differential geometry. Indeed, the gauge potentials represent the generators of the local symmetry group and couple to the respective conserved Noether currents, which act as sources of gravity. In practical terms, the identification of the appropriate gauge field potentials comes from the requirement of covariance of $D\Psi$.

In PGTG the tetrads and the spin connection 1-forms are the gauge potentials, associated with translations, $T(4)$, and Lorentz rotations, $SO(1,3)$, respectively. Torsion and the curvature 2-forms are the respective field strengths. Torsion can be decomposed into 3 irreducible parts $T^a = T^a_{(1)} + T^a_{(2)} + T^a_{(3)}$, made of a tensor part with 16 independent components, a vector part and an axial (pseudo) vector, both with 4 independent components[10]. As for the curvature, it has 36 independent components which can be decomposed into 6 irreducible parts: Weyl (10), Paircom (9), Ricsymf (9), Ricanti (6), scalar (1), and pseudoscalar (1). In addition, there are 6 generators in the Lorentz group with 6 potentials ($\Gamma^{ab} = -\Gamma^{ba}$) and 6 spin (Noether) currents ($s^{ab} = -s^{ba}$). Analogously, there are 4 generators in the group of spacetime translations, which entail 4 gauge potentials θ^a and 4 conserved Noether currents τ^a. By constructing the gravitational Lagrangian with the curvature and torsion invariants, the potentials are coupled to the Noether currents via $24 + 16 = 40$ second order field equations.

Let us now build the different contributions of the matter and gravity fields to the PGTG. For the former, we consider a matter Lagrangian $\mathcal{L}_m = \mathcal{L}_m(g_{ab}, \theta^c, D\Psi)$ such that the covariant derivative with respect to the RC connection, $D\Psi$, allows it to be invariant under local Poincaré spacetime transformations. There are two classes of Noether conserved currents: the canonical energy-momentum tensor density, which is equivalent to the dynamical tetrad energy-momentum density, $\tau_a \equiv \delta L_m / \delta\theta^a$; and the canonical spin density, which is equivalent to the dynamical spin density, $s_{ab} \equiv \delta L_m / \delta\Gamma^{ab}$. These currents couple to the gravitational potentials, acting as sources of gravity, and obey generalized conservation equations. As for the gravity sector in the action, it is constructed with the gauge-invariant gravitational field strengths in the kinetic part associated with the dynamics of the gravitational degrees of freedom.

The total Lagrangian density thus reads

$$\mathcal{L} = \mathcal{L}_G(g_{ab}, \theta^a, T^a, R^{ab}) + \mathcal{L}_m(g_{ab}, \theta^a, D\Psi) \,. \tag{50}$$

By varying this action with respect to the gauge fields of gravity (θ^a, Γ^{ab}) and the matter fields Ψ, we get the corresponding field equations. For the fermionic matter fields, the variational principle $\delta L_m / \delta\Psi = 0$ leads to a generalization of the Dirac equation. As for the bosonic sector (gravity), the inhomogeneous Yang-Mills equations in PGTG are

$$DH_a - \pi_a = \tau_a, \qquad DH_{ab} - \varsigma_{ab} = s_{ab}, \tag{51}$$

where $H_a = -\partial\mathcal{L}_G / \partial T^a$ and $H_{ab} = -\partial\mathcal{L}_G / \partial R^{ab}$ are the 2-form excitations (field momenta) associated to torsion and curvature, and $\tau_a \equiv \delta\mathcal{L}_m / \delta\theta^a$ and $s_{ab} \equiv \delta\mathcal{L}_m / \delta\Gamma^{ab}$ are the 3-form canonical energy-momentum and spin currents. The 3-forms π_a and ς_{ab} can be interpreted as the energy-momentum and spin of the gravitational gauge fields, respectively, defined as

$$\pi_a \equiv e_a \lrcorner L_G + (e_a \lrcorner T^b) \wedge H_b + (e_a \lrcorner R^{cd}) \wedge H_{cd}, \tag{52}$$

$$\varsigma_{ab} \equiv -\theta_{[a} \wedge H_{b]} \,. \tag{53}$$

For a given theory, one only needs to compute the excitations, the source currents and the gravitational currents from the Lagrangian density (50) and substitute directly in the inhomogeneous Equation (51). Note that in this formalism of exterior forms these field equations are completely metric-free, fully general and coordinate-free, with well defined gravitational energy-momentum and spin currents. The PGTG have two sets of Bianchi identities, previously introduced as $DR^a_b = 0$ and $DT^c = R^c_d \wedge \theta^d$, which are intrinsic to the geometrical structure of RC spacetime. Via the field equations, these can be related to the generalized conservation equations for the energy-momentum and the spin currents.

[10] In the minimal coupling to fermions, only the axial vector torsion is involved.

Regarding the constitutive relations in these theories, the 2-form excitations expressed in terms of the field strengths (torsion and curvature), $H_a = H_a(T^c, R^{bd})$ and $H_{ab} = H_{ab}(T^c, R^{bd})$, represent two sets of constitutive relations and are implicit in the Lagrangian formulation. These excitations are in exact analogy to the canonically conjugate momenta of classical mechanics, while torsion and curvature are the generalized velocities for the gravitational degrees of freedom represented by the gauge potentials. These constitutive relations in vacuum can be interpreted as describing the gravitational propagation properties of the spacetime physical manifold. It is via these relations that the conformally invariant part of the metric is introduced, via the Hodge star operator for instance. Moreover, in these relations the coupling constants of the theory are required in order to adequately convert the dimensions of the field strengths (field velocities) to the excitations (field momenta). As constitutive relations for the spacetime vacuum itself, one can postulate that such coupling constants characterize physical properties of the spacetime manifold endowed with gravitational geometrodynamics. This hypothesis is in clear analogy to the similar interpretation for the electromagnetic properties entering in the corresponding constitutive relations (see Reference [38] for further details).

4.3.1. Quadratic Poincaré Gauge Gravity

PGTG with Lagrangians quadratic in the curvature and torsion invariants have been investigated within cosmology, gravitational waves, and spherical solutions, see for example, References [45–49]. The PGTG class is fundamental given the importance of the Poincaré symmetries in relativistic field theories, and the most general quadratic Lagrangian (*à la* Yang-Mills) contains parity breaking terms induced by the richer RC geometry with curvature and torsion [45]. It can be written as

$$\begin{aligned} L &= \frac{1}{2\kappa^2}\Big[(a_0\eta_{ab} + \bar{a}_0\theta_a \wedge \theta_b) \wedge R^{ab} - 2\Lambda\eta - T^a \wedge \sum_{I=1}^{3}\Big(a_I(\star T_a^{(I)}) + \bar{a}_I T_a^{(I)}\Big)\Big] \\ &\quad -\frac{1}{2\rho}R^{ab} \wedge \sum_{I=1}^{6}\Big(b_I(\star R_{ab}^{(I)}) + \bar{b}_I R_{ab}^{(I)}\Big)\,. \end{aligned} \tag{54}$$

The first term in the first line corresponds to the ECSK theory plus the Holst term $\sim (\theta_a \wedge \theta_b) \wedge R^{ab}$, where $\eta_{ab} \equiv e_b \lrcorner \eta_a = \star(\theta_a \wedge \theta_b)$[11]. The second term corresponds to a cosmological constant. The second line contains the terms quadratic in the torsion field strength and the index $I = 1,2,3$ runs over the three irreducible pieces of torsion. In the third line we have the curvature quadratic terms and the index $I = 1,...,6$ runs over the six irreducible pieces of curvature. The free parameters include the $2+6+12 = 20$ (a,b) coefficients, plus the cosmological constant and the sometimes called "strong gravity" parameter ρ. In this Lagrangian all terms with the coefficients with a bar $\bar{a}_0, \bar{a}_I, \bar{b}_I$ break the symmetry under parity transformations. For specific choices and assumptions this Lagrangian includes, for instance, GR itself, the teleparallel equivalent to GR, or the ECSK theory. In the latter, Dirac fermions have axial-axial contact interactions (the Hehl-Data term) with a repulsive character, while in general quadratic PGTG this contact spin-spin interaction is generalized by predicting a propagating interaction. In particular, intermediating gauge bosons with spins $s = 0,1,2$ are predicted, which correspond to massive or massless scalar, vector[12] and tensor propagating modes, respectively. In these GW fields there are odd parity (parity breaking) modes which could manifest themselves as signatures of chirality in the GW cosmological backgrounds from the early Universe[13]. Let us also note that in PGTG it is possible to identify ghost-free Lagrangians which can also be quantized [50,51].

---

[11] Here $\eta_a = e_a \lrcorner \eta = \star\theta_a$ is a 3-form and $\eta = \star 1$ is the natural volume 4-form.
[12] This is actually a torsion axial vector which couples to elementary particles.
[13] For that, one needs non-planar detectors, 3-point correlation functions analysis and sufficient signal/noise ratio, besides a clear distinction from other possible GW sources with a similar power spectrum signature.

4.3.2. The Teleparallel Equivalent of GR

Choosing the Weitzenböck spacetime geometry for PGTG, which implies keeping only torsion while setting both curvature and non-metricity to zero, it is possible to formulate a Lagrangian quadratic in torsion subject to some specific restrictions (such as an appropriate choice for the Weitzenböck connection). This theory can be formulated from a translational gauge theory of gravity perspective. Heuristically, one gets in this case

$$D_\alpha T_c{}^{\alpha\gamma} + (...) \sim \kappa^2 \tau_c{}^{\gamma} \,, \tag{55}$$

or, in terms of the tetrads

$$\Box\theta^a{}_\mu + (...) \sim \kappa^2 \tau^a_\mu \,, \tag{56}$$

where the missing terms on the left-hand side are non-linear terms, and $\Box$ is a d'Alembertian operator. The last equation resembles its GR counterpart, $\Box g_{\mu\nu} + ... \sim \kappa^2 T_{\mu\nu}$, and it turns out that both equations yield the same gravitational phenomenology for matter described by fundamental scalar or Maxwell fields, where the canonical $\tau_{\mu\nu}$ and the dynamical (Einstein-Hilbert) $T_{\mu\nu}$ energy-momentum tensors coincide. For fermions the theories are not fully equivalent. It is interesting to point out that the quadratic (in torsion) Lagrangian of this teleparallel equivalent of GR (TEGR) is locally Lorentz invariant and equivalent to the Hilbert-Einstein Lagrangian [39], but if one wants to formulate GR as a gauge theory then one must gauge the translational group instead of the Lorentz group. This provides yet another motivation to go beyond GR using in this case the teleparallel formulation of gravity to explore exact solutions, post-Newtonian limit, gravitational waves, and so forth, References [52–56], since it is plausible to consider the whole Poincaré symmetries in Nature to be valid, not only the translational group.

4.3.3. Einstein-Cartan-Sciama-Kibble Gravity

In the formalism of exterior forms, the ECSK Lagrangian can be written as

$$\mathcal{L} = \frac{1}{2\kappa^2}\eta_{ab} \wedge R^{ab} \,. \tag{57}$$

The field equations are then obtained by varying this action with respect to the tetrads and the (Lorentzian) spin connection. In the more common tensor formalism this Lagrangian corresponds to the linear Lagrangian in the curvature scalar, yielding the action

$$S_{\text{EC}} = \frac{1}{2\kappa^2}\int d^4x\sqrt{-g}R(\Gamma) + \int d^4x\sqrt{-g}\,\mathcal{L}_m \,. \tag{58}$$

In this expression $\kappa^2 = 8\pi G$ with G Newton's coupling constant, and g is the determinant of the spacetime metric $g_{\mu\nu}$. In the RC spacetime the curvature scalar $R(\Gamma)$ defined via (28) includes terms quadratic in torsion, while the matter Lagrangian, $\mathcal{L}_m = \mathcal{L}_m(g_{\mu\nu}, \Gamma, \Psi_m)$ depends on the metric and the matter fields, Ψ_m, and also on the contortion (i.e., on torsion) via the covariant derivatives.

The Cartan equations can be obtained by varying the action (58) with respect to the contortion tensor $K^\alpha{}_{\beta\gamma}$ (or, alternatively, with respect to the spin connection), and the corresponding result can be written as

$$T^\alpha{}_{\beta\gamma} + T_\gamma\delta^\alpha_\beta - T_\beta\delta^\alpha_\gamma = \kappa^2 s^\alpha{}_{\beta\gamma} \,, \tag{59}$$

where

$$s^{\gamma\alpha\beta} \equiv \frac{\delta\mathcal{L}_m}{\delta K_{\alpha\beta\gamma}} \,, \tag{60}$$

is the spin density tensor, while $T_\beta \equiv T^\gamma{}_{\beta\gamma}$ is the torsion (trace) vector. Cartan's Equation (59) imply that torsion is related to the spin density of matter fields via linear and algebraic relations and, therefore, in the absence of spin (as in vacuum) torsion vanishes. On the other hand, the variation of the action (58) with respect to the spacetime metric $g_{\mu\nu}$ (or, alternatively, with respect to the tetrads) yields the generalized Einstein equations, which, can be written as $G_{\mu\nu} = \kappa^2 \tau_{\mu\nu}$, where $G_{\mu\nu}$ is the Einstein tensor of the RC geometry and $\tau_{\mu\nu}$ is the canonical energy-momentum. These equations can also be conveniently written as

$$\tilde{G}_{\mu\nu} = \kappa^2 T^{\text{eff}}_{\mu\nu} , \tag{61}$$

where $\tilde{G}_{\mu\nu}$ is the Einstein tensor computed with the Levi-Civita connection, while the effective stress-energy tensor

$$T^{\text{eff}}_{\mu\nu} = T_{\mu\nu} + U_{\mu\nu} = -\frac{2}{\sqrt{-g}} \frac{\delta(\sqrt{-g}\mathcal{L}^{\text{eff}}_m)}{\delta g^{\mu\nu}} , \tag{62}$$

includes the (dynamical) metric energy-momentum tensor of the matter fields, $T_{\mu\nu} = -\frac{2}{\sqrt{-g}}\frac{\delta(\sqrt{-g}\mathcal{L}_m)}{\delta g^{\mu\nu}}$, and the additional piece $U_{\mu\nu} = -\frac{2}{\sqrt{-g}}\frac{\delta(\sqrt{-g}C)}{\delta g^{\mu\nu}}$, with $C \equiv -\frac{1}{2\kappa^2}\left(K^{\gamma}{}_{\beta\gamma}K^{\alpha\beta}{}_{\alpha} + K^{\alpha\lambda\beta}K_{\lambda\beta\alpha}\right)$, which contains corrections which are quadratic in torsion $U \sim \kappa^{-2}T^2$. This piece can also be expressed in terms of the spin tensor using Cartan's Equation (59), i.e., $U \sim \kappa^2 s^2$. Note also that, in general, torsion also contributes to the energy-momentum tensor $\tau_{\mu\nu}$, since the covariant derivatives present in the kinetic part of $\mathcal{L}_m$ introduce new terms depending on torsion via either minimal or non-minimal couplings. Since $U \sim \kappa^2 s^2$, Equation (61) defines a typical density, known as Cartan's density, and given by $\rho_C \sim 10^{54}$ g/cm^3 (if one considers nuclear matter[14]). Therefore, in principle, ECSK theory can only introduce significant physical effects in environments of very large spin densities, which might arise in the early universe or in the innermost regions of black holes.

To conclude this part, let us mention that, once the coupling to fermions is considered, the generalized Dirac (Hehl-Data) equation for spinors in RC spacetime, coupled to the ECSK gravity, is cubic in the spinors and includes a torsion induced spin-spin (axial-axial) contact interaction. This type of interactions have been searched for in particle physics including studies at HERA, LEP and Tevatron in electron-proton scattering [57,58].

4.4. Quadratic Gauge Gravity Models in Metric-Affine Gravity

One can generalize the PGTG by considering the affine group as the gauge symmetry group for gravity. If this is performed *à la* la Yang-Mills, then one gets quadratic models as in quadratic PGTG, where the quadratic terms are sometimes referred to as (hypothetical) "strong gravity" terms. This idea has been recovered from time to time, as in Yang [59], in the tensor dominance model [60] or in chromogravity [61]. Depending on the choice of the Lagrangian, the strong gravity (bosonic sector) can be very massive or massless. In some respect these gauge bosonic gravity fields are similar to the Yang-Mills bosons, and if they are massive it is typically assumed that the masses are of the order of the Planck mass or even above. As in quadratic PGTG, in quadratic metric-affine models there are intermediate gauge bosons with spins ($s = 0, 1, 2$) corresponding to scalar, vector or tensor modes (massive or massless). In this respect, the quadratic model in Equation (54) for the PGTG can be extended to metric-affine gravity by including terms with Q_{ab} and $R_{(ab)}$ which are both zero in PGTG, in accordance with the choice of a Lorentzian connection and, therefore, of zero non-metricity (see References [62–64]).

---

14 In cosmological applications the critical density can be written as $\rho_{\text{crit}} \sim m/\lambda_{\text{Comp}} l^2_{\text{Pl}}$, where l_{Pl} and λ_{Comp} are Planck's length and Compton wavelength, respectively. For electrons we get $\rho_{\text{crit}} \sim 10^{52}$ g/cm^3, corresponding to $T_{\text{crit}} \sim 10^{24} K$ and around $t \sim 10^{-34}$ s after the Big Bang.

As an example of a quadratic metric-affine model, one can consider a Lagrangian density (for the bosonic-gravity sector) of the form (schematically)

$$\mathcal{L}_{MAG} \sim \frac{1}{\kappa^2}\left(R + T^2 + TQ + Q^2\right) + \frac{1}{\rho}\left(W^2 + Z^2\right), \tag{63}$$

where $W_{ab} \equiv R_{[ab]}$, and $Z_{ab} \equiv R_{(ab)}$ are known as the "rotational curvature" and the "strain curvature", respectively. The terms proportional to the coupling constant ρ are referred to as strong gravity terms, in contrast to the terms that are proportional to the "weak" gravity coupling constant κ^2. Note that in this model the connection is non-Lorentzian and the non-metricity 1-form (represented by Q) is non-vanishing. The bosonic sector of metric-affine gravity was analyzed, for instance, in References [62–65], while the fermionic part is more delicate (see for instance References [66–69]). In this respect, there is no finite dimensional spinor representation of the $GL(4,\Re)$ group, which leads to the introduction of the "world spinors" (with infinite components) of Ne'eman, and to the corresponding generalization of Dirac equation. The world spinor formalism is related to Regge trajectories, which are themselves related to spin-2 excitations of hadrons (see Reference [70]).

For an interesting review on exact solutions in MAG see Reference [71]. Further research in this context involves cosmological scenarios [72], and one should also mention exact spherically symmetric solutions with $Q \sim 1/r^d$ [73], which suggests the existence of massless modes.

4.5. Probing Non-Riemannian Geometry with Test Matter

It is known that torsion can give rise to precession effects in systems with intrinsic spin, for example, elementary particles such as the electron, or baryons such as the neutron [74,75]. This is a model-independent result which can be obtained from a (WKB) semi-classical approximation of the Dirac fermionic dynamics in a RC spacetime. In principle this prediction can be used to distinguish between the spacetime paradigms of GR and TEGR. If v is the polarization vector of the (intrinsic) spin then one can deduce the simple expression

$$\dot{v} = 3\check{t}v\,, \tag{64}$$

where the axial vector $\check{t}$ is given by $\check{t}^\alpha \equiv -\epsilon^{\alpha\beta\gamma\delta}T_{\beta\gamma\delta}$. In this sense it is plausible that experiments similar to the Gravity Probe-B but using gyroscopes with intrinsic (macroscopic) spin can be used to constrain or detect the effects of the (hypothetical) torsion around the Earth [76]. There have been quite a number of studies on spin precession effects induced by torsion (see also Reference [77]). On the other hand, Lammerzahl have set experimental limits for detecting torsion ($|T| \sim 10^{-15}m^{-1}$) using Hughes-Drever (spectroscopic) type of experiments [78].

Moreover, one can predict torsion effects on the energy levels of quantum systems. From the Dirac Lagrangian of a fermion minimally coupled to the background RC geometry

$$\mathcal{L}_{\text{Dirac}} = \frac{i\hbar}{2}\left(\bar{\psi}\gamma^\mu D_\mu\psi - (D_\mu\bar{\psi})\gamma^\mu\psi\right) - m\bar{\psi}\psi\,, \tag{65}$$

(for spinors ψ and their adjoints $\bar{\psi}$) one deduces the Dirac equation

$$i\hbar\gamma^\mu\tilde{D}_\mu\psi - m\psi = -\frac{3\hbar}{2}\check{T}^\lambda\gamma_\lambda\gamma^5\psi\,, \tag{66}$$

which can be analyzed in the flatness limit. If we assume a static axial torsion vector $\check{T}^\lambda$ along the x^3-direction, then we get the time-independent wave equation

$$-i\hbar\gamma^k\partial_k\psi + \left(m - \frac{3\hbar}{2}\check{T}^3\gamma_3\gamma^5\right)\psi(\vec{r}) = \gamma^0 E\psi(\vec{r})\,. \tag{67}$$

From this equation one obtains two possible values for the energy levels, depending on whether the fermionic spin is aligned or anti-aligned with respect to the background (axial) torsion, that is [79],

$$E^2 = p^2 + \left(m \pm \frac{3\hbar}{2}\check{T}^3\right)^2 . \tag{68}$$

Similarly, if non-minimal couplings between the torsion trace vector and Dirac axial vector and/or between torsion axial vector and Dirac vector $\bar{\psi}\gamma^\mu\psi$, are present, then the parity symmetry is broken and the corresponding energy level corrections due to torsion will contain the signatures of those parity breaking interactions. Therefore, tests with advanced spectrographs might be able to probe torsion effects on quantum systems. Also, the previously mentioned spin-spin contact interactions of the ECSK theory, or the propagating spin-spin interactions mediated by gravitational gauge ($s = 0, 1, 2$) bosons might be tested/constrained in laboratory experiments and cosmological GW probes.

Regarding non-metricity, the gauge approach to gravity clearly shows that the hypermomentum currents, such as the dilatations or the shear currents, couple to the trace vector part and the shear part of the non-metricity, respectively. This coupling is evident since, as we have seen, the connection couples to hypermomentum and a non-Lorentzian connection implies non-metricity. Further developments have shown that, if torsion can be measured by the spin precession of test matter with intrinsic spin, then the non-metricity of spacetime can be measured by pulsations (mass quadrupole excitations) of test matter with (non-trivial) hypermomentum currents. In order to be "sensitive" to non-Riemann geometries, test matter should carry dilatation, shear, or spin currents, whether macroscopic or at the level of fundamental fields/particles. In the latter case, the Regge trajectories provide an adequate mathematical illustration of test matter as Ne'eman's world spinors with shear.

Let us also point out that Obukhov and Puetzfeld [80] have derived the equation of motion for matter fields in metric-affine gravity. By making use of the Bianchi identities one can arrive at the following expression for the translational Noether current:

$$\tilde{D}\left[\tau_a + \Delta^{bc}\left(e_a \lrcorner N_{bc}\right)\right] + \Delta^{bc} \wedge \left(\pounds_{e_a} N_{bc}\right) = s^{bc} \wedge \left(e_a \lrcorner \tilde{R_{cb}}\right), \tag{69}$$

where, as usual, the tilde refers to quantities defined in the Riemann geometry, while τ_a, Δ^{bc}, s^{bc} are the canonical energy-momentum, hypermomentum current, and spin current, respectively, and $N_{bc} = \Gamma_{bc} - \tilde{\Gamma}_{bc}$ gives the non-Riemannian piece of the connection 1-form, that is, the distortion 1-form. Note that in the right-hand side of this equation we can identify the Mathisson-Papapetrou force density for matter with spin. In standard GR, one obtains $\tilde{D}\tau_a = 0$, which gives the geodesic equation for spinless matter with energy-momentum, while for $N_{bc} = 0$ we get $\tilde{D}\tau_a = s^{bc} \wedge (e_a \lrcorner \tilde{R}_{cb})$, which is the Mathisson-Papapetrou equation in GR for matter with spin. In this equation, if matter has neither (intrinsic) spin, nor dilatation/shear currents, then it follows the Riemannian (extremal length) geodesics, regardless of the geometry of spacetime or of the form of the Lagrangian in metric-affine geometry.

4.6. Metric-Affine Geometry and Lorentz Symmetry Breaking

Regarding the analogous law for the $GL(4, \Re)$ Noether current we have

$$D\Delta^a_{\ b} + \theta^a \wedge \tau_b = \tau^a_{\ b} \, . \tag{70}$$

In PGTG the connection is Lorentzian $\Gamma^{ab} = -\Gamma^{ba}$, while in the WGTG the trace part of the connection $\frac{1}{4}g^{ab}\Gamma^c_{\ c}$ is also non-vanishing. A fully independent connection in metric-affine theory is given by the expression in (6). It is this linear connection that couples to the (intrinsic) hypermomentum current (see Equation (25)). In particular, the Lorentzian connection couples to spin s_{ab} carrying $SO(1,3)$ charges, while the trace part couples to the dilatation current $\Delta^c_{\ c}$, and the shear part of the connection (see Equation (26)) couples to the shear current $\bar{\Delta}_{ab}$ that carries $SL(4,\Re)/SO(1,3)$ (intrinsic) shear charges. The shear current seems to be related to the Regge trajectories [69] and these represent spin-2 excitations of hadrons with the same internal quantum numbers. The relation between the Regge trajectories (which can be described by the group $SL(3,\Re)$ and this can be embedded in the $SL(4,\Re)$ one) and the hypermomentum shear charges, remains an open question under study and its validity seems to point to a quite remarkable and promising connection between the strong interaction of hadrons and spacetime post-Riemann geometry. The shear charge is actually a measure of the breaking of Lorentz invariance. The bottom line is that, in order to get Lorentz symmetry breaking, one does not require the introduction of extra particle degrees of freedom, but this can be obtained solely via the geometrical structures of spacetime, namely a non-Lorentzian connection. The presence of a non-Lorentzian connection implies the non-vanishing non-metricity and the non-vanishing strain curvature $R_{(ab)}$.

4.7. A Word on the Formulations of GR

The (canonical) metric formulation of GR requires a pseudo-Riemannian manifold with a symmetric, $\Gamma^{\alpha}_{\mu\nu} = \Gamma^{\alpha}_{\nu\mu}$, and metric-compatible, $\nabla^{\Gamma}_{\alpha} g_{\mu\nu} = 0$, affine connection (the Levi-Civita one). However, it is well known nowadays that the teleparallel equivalent of GR, formulated in the Weitzenböck spacetime, yields a dynamically equivalent theory to metric GR, with exactly the same predictions [39,81–83]. Besides this approach from a translational gauge principle under specific assumptions, there is yet another formulation equivalent to GR based on zero curvature and torsion but non-zero non-metricity, called symmetric teleparallel gravity, whose properties have begun to be unravelled very recently [84–88].

Since these are, to some extent, equivalent gravitational models under different spacetime paradigms, one may ask if there is any guiding principle which could determine what spacetime geometry and degrees of freedom can represent gravity at its most fundamental level. The application of the gauge approach to gravity shows clearly that GR can be formulated as a translational gauge theory and, therefore, lives on a subset of the RC spacetime. On the other hand the generalization of the translational symmetry to the Poincaré symmetries points towards the direction of the PGTG formulated in the RC spacetime geometry. It is relevant to underline (once again) that the three mentioned approaches to GR have different assumptions regarding the spacetime geometrical paradigm, but their equivalence breaks as soon as one considers matter with spin (spinors). Indeed, if one could measure the different effects of non-Riemannian geometries upon matter, one might be able to distinguish between these spacetime paradigms.

5. Discussion and Future Outlook

5.1. Spacetime Paradigms

In this work we explored geometrical methods (post-Riemann spacetime geometries and Cartan's exterior calculus of forms) and symmetry principles in the gauge approach to gravity, and how these topics might point towards a new perspective over the spacetime paradigm. We also briefly considered the pre-metric formulation of classical electrodynamics. In this broad perspective, the conformal-causal structure is argued to be more fundamental than the metric structure (the primacy of the conformal geometry) and the absoluteness of the spacetime metric is then abandoned at the fundamental level. We also established the analogies between the pre-metric canonical formulation of gauge theories of gravity and the pre-metric equations and mathematical objects of general Yang-Mills fields. The theoretical formulation of the Lagrangians in these theories (gravity and Yang-Mills) implicitly presuppose the assumption of the specific form for the so-called constitutive relations between the field strengths (the generalized field velocities) and the excitations (conjugate field momenta). These relations can be interpreted as constitutive relations for the spacetime itself. Moreover, in this discussion we highlighted the hypothesis that the physical constants or coupling parameters that enter in such relations reflect physical properties of the spacetime manifold, itself regarded as a mathematical object that represents a truly physical system.

By endowing the classical spacetime with physical properties, the concept of classical vacuum with properties such as electric permitivity, magnetic permeability, and so forth, becomes somehow disposable or simply dual to the very notion of a physical spacetime. Moreover, the properties of this physical spacetime might change from point to point and this scenario fits well within the scalar-tensor (Brans-Dicke, etc.), vector-tensor or tensor-tensor extensions to GR. The idea that these properties of spacetime can be described by fields can have implications to spacetime symmetry considerations, i.e., the invariance of the physics under groups of spacetime coordinate transformations, and also a link to the Mach's ideas and the breaking of spacetime (metric) absoluteness. Therefore, this scenario of a physical spacetime with non-Riemann geometry and physical properties described by fields fits naturally very well in the assumption of the primacy of the conformal-causal structure. This means that the so-called constants can change from place to place in space (non-homogeneity) and with spatial direction (anisotropy) and still preserve the local conformal-causal symmetry and, therefore, by extension the causal structure. Consequently, what is assumed to be physically relevant are the conformally-invariant properties. In other words, one must seek for an extended spacetime manifold with post-Riemann geometries, physical properties, and a fundamental conformal symmetry.

Keeping this trail of thought, one arrives at the conclusion that the intrinsic physical properties of spacetime include energy-momentum, hypermomentum (including spin), electromagnetic and thermodynamical properties associated to gravity and to the spacetime. The latter actually suggests the existence of microscopic degrees of freedom, and if these are assumed to constitute a numerable set, then a consistent picture should go beyond the classical manifold and incorporate some degree of discreteness or gravity quantization. Therefore, a physical spacetime manifold with locally invariant conformal-casual structure, with intrinsic physical properties (electromagnetic, thermodynamical, etc.) and possibly a quantum nature seems the natural hypothesis to address the unification of spacetime, matter fields and the quantum and classical vacuum.

5.2. *Perspectives on Unification Methods in Fundamental Field Theories*

Let us now explore in more detail the hypothesis stated above, as well as its relation to various aspects of unification in physics. We shall put here our emphasis upon the conceptual background motivated by mathematical considerations and, in particular, from geometrical methods and symmetry principles.

Firstly, the geometrical methods in the pre-metric formalism of electromagnetism, Yang-Mills and gravity field equations, using the calculus of exterior forms, together with the corresponding constitutive relations (that can be interpreted as (spacetime) constitutive relations, suggest the following: (i) the primacy of the conformal-causal structure (the conformally-invariant part of the metric) over the full metric structure, and therefore, (ii) the assumption of absoluteness of the spacetime metric ("absolute spacetime") is abandoned at the fundamental level. Also one postulates that the fundamental coupling constants entering in the (vacuum) constitutive relations represent physical properties of spacetime, not necessarily spatially homogeneous and isotropic (constants), while respecting the local conformal symmetries. As a consequence we emphasize again here the first unification: the identification of the physical classical vacuum with physical classical spacetime.

Secondly, from gauge symmetries in gravitation and post-Riemann geometries we find the need to consider a classical spacetime with general metric-affine geometry, namely, having non-vanishing curvature, torsion and non-metricity. Indeed, the gauge approach to gravity requires the existence of post-Riemann geometries associated to gravitational phenomena. This motivates the search for its signatures via astrophysical and cosmological effects, including GW probes and effects in particle physics. On the other hand, the idea that the spacetime metric might not be fundamental is rooted on the mathematical consistency of gauge theories of gravity which links the symmetries of physics under coordinate transformations to the geometrical paradigm of spacetime. The theory identifies clearly, for each specific gauge group, the underlying spacetime geometry, the Noether currents (which are the sources of the gravitational field) and the gauge potentials (geometrical degrees of freedom) and their field strengths. In this formalism the spacetime metric is not a fundamental field. This is clearly seen in the language of forms, where an explicit pre-metric approach to the field equations is completely general, coordinate-free and covariant.

The metric structure can be assumed *a posteriori*, in particular, via the constitutive relations (which relate the field strengths or field velocities, that is, curvature and torsion, to the canonically conjugate field momenta). More specifically, using these relations, via the Hodge star operation, only the conformally invariant part of the metric structure is involved in the coupling between the field strengths and the Noether sources, rather than the full metric structure. Therefore, symmetry principles and geometrical methods in gauge theories of gravity suggest again that the (full) metric structure is not fundamental, and in models with larger symmetry groups beyond the Poincaré group the paradigm of spacetime (metric) absoluteness is not valid, that is, the metric changes under specific coordinate transformations between local observers. If non-metricity is present, the connection is non-Lorentzian and, as a consequence, one gets the breaking of Lorentz symmetry. It is this type of geometry that is associated to the conformal structure that emerges from linear electrodynamics [89].

Regarding unified symmetry groups and symmetry breaking, it is known that general symmetry groups can be broken into smaller groups within phase transitions. In principle, these are expected to have occurred in the early Universe in clear analogy to the standard model (and beyond the standard model) of interactions, leading to first order phase transitions and the generation of GW emission in the form of a stochastic GW background. This GW signature of cosmological origin (which might include polarization features due to parity breaking gravitational physics of general quadratic Yang-Mills gauge theories of gravity) might be detectable by future missions such as LISA [90]. Of particular interest is the conformal group of CGTG, that may be broken into the Poincaré group. This necessarily includes the breaking of scale invariance and the emergence of natural physical scales and the corresponding (Lorentz invariant) fundamental constants. This presupposes the existence of a scale-invariant cosmological epoch where the properties of the physical spacetime obey perfect conformal symmetry and its geometry, and also a transition into a broken symmetry phase where the spacetime metric appears to have a (local) absolute nature. Whether there might be other extreme physical regimes (such as the innermost regions of ultra-compact objects and black holes) where scale invariance, or even the full conformal symmetry, is recovered, remains an open question.

Another suggestion is that curvature, torsion and non-metricity might be inter-convertible. This seems appropriate from the point of view of the generalized Bianchi identities, (29) and (22), of metric-affine geometries relating the field strengths of gauge theories of gravity (curvature, torsion and non-metricity). This is somewhat analogous to the magnetic-electric inter-conversion of electromagnetism. Recall that the $dF = 0$ is a Bianchi identity giving magnetic flux conservation and Faraday law. These post-Riemann Bianchi identities are implicit to the spacetime geometrical structures and express some sort of gravitational flux conservation, and are compatible with the Noether currents conservations of the gauge approach to gravity. These relations point towards this notion of inter-conversion, which could open new avenues for the study of gravitational phenomena in extreme regimes.

Moreover, this hypothesis resembles the Weyl-Ricci conversion (the Weyl conjecture) in cosmology, within GR. According to it, the Ricci curvature dominates completely the very early Universe and the Weyl curvature dominates the late-Universe, in such a way that the Ricci part of the Riemann tensor is converted (transformed) into the Weyl part as the Universe expands, and forms gravitationally bound structures, asymptotically dominated by black holes. Similarly, the dynamical transformations (conversions) between the three parts inside the full curvature of MAG can be compatible to the (generalized) Bianchi identities involving the full curvature. This interpretation is further reinforced by the existence of formal maps between "equivalent" descriptions of GR in the spacetimes with torsion, curvature or non-metricity alone (and also between the generalizations of GR in the respective spacetime paradigms). Although under specific formal conditions the phenomenology of these theories might be equivalent, the spacetime paradigms are obviously different and the minimal/non-minimal couplings of fermionic fields to these geometries can break the equivalences. The idea that the Poincaré or the $GL(4, \Re)$ groups should be fundamental points towards generalizations of GR in any of these quasi-equivalent descriptions, via PGTG or MAG, and in this context the mathematical structure of the gauge approach seems to be compatible with the interpretation that curvature, torsion and non-metricity can indeed be inter-convertible.

In addition to the topics discussed above it is worth mentioning that the gauge approach is transversal to both gravity and Yang-Mills fields of the standard model of particles and interactions. Although in the first case one speaks of external or spacetime symmetries, while in the second case the spinorial character of elementary particles/fields requires internal symmetries, in both cases there is a deep relation between group theory (symmetries) and the dynamics of fundamental fields via geometrical methods. The gauge formalism thus highlights its potential for bridging spacetime geometry and gravity with the other fundamental interactions. The challenges are precisely those of how to relate internal spaces and its symmetries with the spacetime geometry and its symmetries in a unified way. In this context, we mentioned briefly the supersymmetric gauge theories of gravity, which somehow aim to achieve that goal by enlarging the gauge formalism and unifying bosons and fermions under the same mathematical structure.

5.3. Final Remarks

The potential of the geometrical methods and symmetry principles described in this work to establish analogies and connections between different interactions clearly supports their vital contribution for unified field theories. It is quite consensual in the community that the road to unification and quantum gravity will inevitably lead to a new spacetime paradigm. Whether this unification will imply some convergence between spacetime and physical fields into a single "physical spacetime-matter' entity' is yet to be seen. It is worth mentioning at such a level the possibility of the discretization of the underlying geometry of the spacetime, which is suggested on several grounds, in particular, on the connection of metric-affine geometries and solid state physics systems with defects in their microstructure [91], or on the thermodynamics of gravitational fields and the entropy of spacetime regions (horizons) in terms of internal microstates. This is linked with the challenge of making compatible the causal/conformal and metric structures of spacetime [92] and the indeterminacy principle of physical fields. Should the latter be merged with spacetime in a single unified entity, then such a manifold would have to be quantized since it needs to include the indeterminacy principle in its intrinsic geometrical nature.

The unified spacetime-matter manifold conjectured above might have the following ingredients: a conformal symmetry at a fundamental level (possibly inside some unified symmetry group as in SUSY-SUGRA); have more than four spacetime dimensions; include complex numbers (so that the internal symmetries of spinors are embedded and intrinsic to the physical unified manifold); have internal and external physical properties such as energy-momentum, hypermomentum, $U(1)$ and $SU(N)$ charges; have electromagnetic (electric permitivity and magnetic permeability) and thermodynamical properties, and include internal degrees of freedom linked by internal $SU(N)$ symmetries as physical degrees of freedom (in analogy to the microscopic sates of macroscopic systems in statistical physics).

To conclude, we add some remarks. First, it seems clear that the fundamental symmetry principles and geometrical methods in the gauge approach to gravity lead to quite remarkable predictions about the nature of spacetime and gravity, which might be tested with astrophysical (compact objects [93]) and cosmological (inflation, late-time evolution [94]) observations. Second, these methods provide insights on the nature of the gravitational currents which stimulate the research on the fundamental nature of matter and its physical conserved properties. Indeed, understanding the hypermomentum currents from a fundamental point of view might lead to remarkable connections between gravity and hadronic physics. Similar comments apply to the appropriate test-matter properties required to probe post-Riemann geometries, which is needed in order to distinguish experimentally between different classical spacetime paradigms. And finally, these methods can provide robust mathematical frameworks for the search of unified field theories, where spacetime, fermions and bosons might be inextricably linked within a common unified physical framework.

Author Contributions: All authors contributed equally to the manuscript. All authors have read and agreed to the published version of the manuscript.

Funding: F.C. is funded by the Fundação para a Ciência e a Tecnologia (FCT, Portugal) doctoral grant No. PD/BD/128017/2016. F.S.N.L. acknowledges support from the FCT Scientific Employment Stimulus contract with reference CEECIND/04057/2017. D.R.-G. is funded by the *Atracción de Talento Investigador* programme of the Comunidad de Madrid (Spain) No. 2018-T1/TIC-10431, and acknowledges further support from the Ministerio de Ciencia, Innovación y Universidades (Spain) project No. PID2019-108485GB-I00/AEI/10.13039/501100011033, the Spanish project No. FIS2017-84440-C2-1-P (MINECO/FEDER, EU), the project PROMETEO/2020/079 (Generalitat Valenciana), and the Edital 006/2018 PRONEX (FAPESQ-PB/CNPQ, Brazil) Grant No. 0015/2019. The authors also acknowledge funding from FCT Projects No. UID/FIS/04434/2020, No. CERN/FIS-PAR/0037/2019 and No. PTDC/FIS-OUT/29048/2017. This article is based upon work from COST Actions CA15117 and CA18108, supported by COST (European Cooperation in Science and Technology).

Acknowledgments: F.C. thanks the hospitality of the Department of Theoretical Physics and IPARCOS of the Complutense University of Madrid, where part of this work was carried out.

Conflicts of Interest: The authors declare no conflict of interest.

References

1. Will, C.M. The Confrontation between General Relativity and Experiment. *Living Rev. Rel.* **2014**, *17*, 4. [CrossRef] [PubMed]
2. Hees, A.; Do, T.; Ghez, A.M.; Martinez, G.D.; Naoz, S.; Becklin, E.E.; Boehle, A.; Chappell, S.; Chu, D.; Dehghanfar, A.; et al. Testing General Relativity with stellar orbits around the supermassive black hole in our Galactic center. *Phys. Rev. Lett.* **2017**, *118*, 211101. [CrossRef] [PubMed]
3. Abbott, B.P.; Abbott, R.; Abbott, T.D.; Abernathy, M.R.; Acernese, F.; Ackley, K.; Adams, C.; Adams, T.; Addesso, P.; Adhikari, R.X.; et al. Observation of Gravitational Waves from a Binary Black Hole Merger. *Phys. Rev. Lett.* **2016**, *116*, 061102. [CrossRef]
4. Abbott, B.P.; Abbott, R.; Abbott, T.D.; Abernathy, M.R.; Acernese, F.; Ackley, K.; Adams, C.; Adams, T.; Addesso, P.; Adhikari, R.X.; et al. GW170817: Observation of Gravitational Waves from a Binary Neutron Star Inspiral. *Phys. Rev. Lett.* **2017**, *119*, 161101. [CrossRef] [PubMed]
5. Abbott, B.P.; Abbott, R.; Abbott, T.D.; Abernathy, M.R.; Acernese, F.; Ackley, K.; Adams, C.; Adams, T.; Addesso, P.; Adhikari, R.X.; et al. Tests of general relativity with GW150914. *Phys. Rev. Lett.* **2016**, *116*, 221101; Erratum in **2018**, *121*, 129902. [CrossRef] [PubMed]
6. Abbott, B.P.; Abbott, R.; Abbott, T.D.; Abernathy, M.R.; Acernese, F.; Ackley, K.; Adams, C.; Adams, T.; Addesso, P.; Adhikari, R.X.; et al. GW170814: A Three-Detector Observation of Gravitational Waves from a Binary Black Hole Coalescence. *Phys. Rev. Lett.* **2017**, *119*, 141101. [CrossRef] [PubMed]
7. Akiyama, K.; Alberdi, A.; Alef, W.; Asada, K.; Azulay, R.; Baczko, A.K.; Ball, D.; Baloković, M.; Barrett, J.; Bintley, D.; et al. First M87 Event Horizon Telescope Results. I. The Shadow of the Supermassive Black Hole. *Astrophys. J.* **2019**, *875*, L1.
8. Bull, P.; Akrami, Y.; Adamek, J.; Baker, T.; Bellini, E.; Beltran Jimenez, J.; Bentivegna, E.; Camera, S.; Clesse, S.; Davis, J.H.; et al. Beyond ΛCDM: Problems, solutions, and the road ahead. *Phys. Dark Univ.* **2016**, *12*, 56–99. [CrossRef]
9. Weyl, H. A New Extension of Relativity Theory. *Ann. Phys.* **1919**, *59*, 101. [CrossRef]
10. Weyl, H. Gravitation and the electron. *Proc. Natl. Acad. Sci. USA* **1929**, *15*, 323. [CrossRef]
11. Yang, C.N.; Mills, R.L. Conservation of Isotopic Spin and Isotopic Gauge Invariance. *Phys. Rev.* **1954**, *96*, 191. [CrossRef]
12. Utiyama, R. Invariant theoretical interpretation of interaction. *Phys. Rev.* **1956**, *101*, 1597. [CrossRef]
13. Sciama, D.W. The Physical structure of general relativity. *Rev. Mod. Phys.* **1964**, *36*, 463; Erratum in **1964**, *36*, 1103. [CrossRef]
14. Kibble, T.W.B. Lorentz invariance and the gravitational field. *J. Math. Phys.* **1961**, *2*, 212–221. [CrossRef]

15. Boehmer, C.G.; Burnett, J. Dark spinors with torsion in cosmology. *Phys. Rev. D* **2008**, *78*, 104001. [CrossRef]
16. Poplawski, N.J. Nonsingular, big-bounce cosmology from spinor-torsion coupling. *Phys. Rev. D* **2012**, *85*, 107502. [CrossRef]
17. Poplawski, N.J. Big bounce from spin and torsion. *Gen. Rel. Grav.* **2012**, *44*, 1007–1014. [CrossRef]
18. Vakili, B.; Jalalzadeh, S. Signature transition in Einstein-Cartan cosmology. *Phys. Lett. B* **2013**, *726*, 28–32. [CrossRef]
19. Bronnikov, K.A.; Galiakhmetov, A.M. Wormholes and black universes without phantom fields in Einstein-Cartan theory. *Phys. Rev. D* **2016**, *94*, 124006. [CrossRef]
20. Ivanov, A.N.; Wellenzohn, M. Einstein–cartan Gravity with Torsion Field Serving as an Origin for the Cosmological Constant or Dark Energy Density. *Astrophys. J.* **2016**, *829*, 47. [CrossRef]
21. Cembranos, J.A.R.; Gigante Valcarcel, J.; Maldonado Torralba, F.J. Singularities and n-dimensional black holes in torsion theories. *JCAP* **2017**, *4*, 021. [CrossRef]
22. Cembranos, J.A.R.; Gigante Valcarcel, J. Extended Reissner–Nordström solutions sourced by dynamical torsion. *Phys. Lett. B* **2018**, *779*, 143–150. [CrossRef]
23. Unger, G.; Popławski, N. Big bounce and closed universe from spin and torsion. *Astrophys. J.* **2019**, *870*, 78. [CrossRef]
24. Blagojevic, M.; Hehl, F.W. *Gauge Theories of Gravitation*; Imperical Collage Press: Londin, UK, 2012. [CrossRef]
25. Blagojevic, M. *Gravitation and Gauge Symmetries*; Institute of Physics Publishing: Bristol, UK, 2002.
26. Ponomarev, V.N.; Barvinsky, A.O.; Obukhov, Y.N. *Gauge Approach and Quantization Methods in Gravity Theory*; Nauka: Moscow, Russia, 2017.
27. Cabral, F.; Lobo, F.S.N.; Rubiera-Garcia, D. Einstein–Cartan–Dirac gravity with $U(1)$ symmetry breaking. *Eur. Phys. J. C* **2019**, *79*, 1023. [CrossRef]
28. Cabral, F.; Lobo, F.S.N.; Rubiera-Garcia, D. Cosmological bounces, cyclic universes, and effective cosmological constant in Einstein-Cartan-Dirac-Maxwell theory. *Phys. Rev. D* **2020**, *102*, 083509. [CrossRef]
29. Cabral, F.; Lobo, F.S.N.; Rubiera-Garcia, D. The cosmological principle in theories with torsion: The case of Einstein-Cartan-Dirac-Maxwell gravity. *JCAP* **2020**, *10*, 057. [CrossRef]
30. Obukhov, Y.N.; Hehl, F.W. General relativity as a special case of Poincaré gauge gravity. *Phys. Rev. D* **2020**, *102*, 044058. [CrossRef]
31. Poplawski, N.J. Gravitation, electromagnetism and cosmological constant in purely affine gravity. *Found. Phys.* **2009**, *39*, 307–330. [CrossRef]
32. Kittel, C. *Introduction to Solid State Physics*, 8th ed.; Wiley: New York, NY, USA, 2005.
33. Hehl, F.W.; Obukhov, Y.N. Spacetime metric from local and linear electrodynamics: A New axiomatic scheme. *Lect. Notes Phys.* **2006**, *702*, 163–187.
34. Hehl, F.W.; Obukhov, Y.N. How does the electromagnetic field couple to gravity, in particular to metric, nonmetricity, torsion, and curvature? *Lect. Notes Phys.* **2001**, *562*, 479–504.
35. Lammerzahl, C.; Hehl, F.W. Riemannian light cone from vanishing birefringence in premetric vacuum electrodynamics. *Phys. Rev. D* **2004**, *70*, 105022. [CrossRef]
36. Itin, Y.; Hehl, F.W. Is the Lorentz signature of the metric of space-time electromagnetic in origin? *Ann. Phys.* **2004**, *312*, 60–83. [CrossRef]
37. Hehl, F.W.; Obukhov, Y.N. To consider the electromagnetic field as fundamental, and the metric only as a subsidiary field. *Found. Phys.* **2005**, *35*, 2007–2025. [CrossRef]
38. Cabral, F.; Lobo, F.S.N. Electrodynamics and Spacetime Geometry: Foundations. *Found. Phys.* **2017**, *47*, 208–228. [CrossRef]
39. Jiménez, J.B.; Heisenberg, L.; Koivisto, T.S. The Geometrical Trinity of Gravity. *Universe* **2019**, *5*, 173. [CrossRef]
40. Bamba, K.; Odintsov, S.D.; Sáez-Gómez, D. Conformal symmetry and accelerating cosmology in teleparallel gravity. *Phys. Rev. D* **2013**, *88*, 084042. [CrossRef]
41. Lord, E.A.; Goswami, P. Gauging the conformal group. *Pramana* **1985**, *25*, 635–640. [CrossRef]
42. MacDowell, S.W.; Mansouri, F. Unified Geometric Theory of Gravity and Supergravity. *Phys. Rev. Lett.* **1977**, *38*, 739. [CrossRef]
43. Stelle, K.S.; West, P.C. Spontaneously Broken De Sitter Symmetry and the Gravitational Holonomy Group. *Phys. Rev. D* **1980**, *21*, 1466. [CrossRef]

44. Berkovits, N. SuperPoincare invariant superstring field theory. *Nucl. Phys. B* **1995**, *450*, 90–102; Erratum in **1996**, *459*, 439–451. [CrossRef]
45. Obukhov, Y.N. Poincaré gauge gravity: An overview. *Int. J. Geom. Meth. Mod. Phys.* **2018**, *15*, 1840005. [CrossRef]
46. Obukhov, Y.N. Gravitational waves in Poincaré gauge gravity theory. *Phys. Rev. D* **2017**, *95*,084028 [CrossRef]
47. Blagojević, M.; Cvetković, B. General Poincaré gauge theory: Hamiltonian structure and particle spectrum. *Phys. Rev. D* **2018**, *98*, 024014. [CrossRef]
48. Blagojević, M.; Cvetković, B. Conformally flat black holes in Poincaré gauge theory. *Phys. Rev. D* **2016**, *93*, 044018. [CrossRef]
49. Cembranos, J.A.R.; Valcarcel, J.G.; Maldonado Torralba, F.J. Non-Geodesic Incompleteness in Poincaré Gauge Gravity. *Entropy* **2019**, *21*, 280. [CrossRef]
50. Sezgin, E.; van Nieuwenhuizen, P. New Ghost Free Gravity Lagrangians with Propagating Torsion. *Phys. Rev. D* **1980**, *21*, 3269. [CrossRef]
51. Kuhfuss, R.; Nitsch, J. Propagating Modes in Gauge Field Theories of Gravity. *Gen. Rel. Grav.* **1986**, *18*, 1207–1227. [CrossRef]
52. Beltrán Jiménez, J.; Heisenberg, L.; Iosifidis, D.; Jiménez-Cano, A.; Koivisto, T.S. General teleparallel quadratic gravity. *Phys. Lett. B* **2020**, *805*, 135422. [CrossRef]
53. Hohmann, M.; Krššák, M.; Pfeifer, C.; Ualikhanova, U. Propagation of gravitational waves in teleparallel gravity theories. *Phys. Rev. D* **2018**, *98*, 124004. [CrossRef]
54. Bahamonde, S.; Böhmer, C.G.; Krššák, M. New classes of modified teleparallel gravity models. *Phys. Lett. B* **2017**, *775*, 37–43. [CrossRef]
55. Bahamonde, S.; Dialektopoulos, K.F.; Hohmann, M.; Levi Said, J. Post-Newtonian limit of Teleparallel Horndeski gravity. *arXiv* **2020**, arXiv:2003.11554.
56. de la Cruz-Dombriz, Á.; Farrugia, G.; Said, J.L.; Gómez, D.S.C. Cosmological bouncing solutions in extended teleparallel gravity theories. *Phys. Rev. D* **2018**, *97*, 104040. [CrossRef]
57. Goy, C. Contact interactions at Hera, LEP and Tevatron. *Eur. Phys. J. C* **2004**, *33*, S776–S778. [CrossRef]
58. H1 Collaboration. Search for new physics in $e^{\pm}q$ contact interactions at HERA. *Phys. Lett. B* **2003**, *568*, 35–47. [CrossRef]
59. Yang, C.N. Integral Formalism for Gauge Fields. *Phys. Rev. Lett.* **1974**, *33*, 445. [CrossRef]
60. Isham, C.J.; Salam, A.; Strathdee, J.A. F-dominance of gravity. *Phys. Rev. D* **1971**, *3*, 867. [CrossRef]
61. Ne'eman, Y.; Sijacki, D. Chromogravity: QCD-induced diffeomorphisms. *Int. J. Mod. Phys. A* **1995**, *10*, 4399–4412. [CrossRef]
62. Hehl, F.W.; Kerlick, G.D.; Von Der Heyde, P. On a New Metric Affine Theory of Gravitation. *Phys. Lett. B* **1976**, *63*, 446–448. [CrossRef]
63. Lord, E.A. The Metric Affine Gravitational Theory as the Gauge Theory of the Affine Group. *Phys. Lett. A* **1978**, *65*, 1–4. [CrossRef]
64. Hehl, F.W.; McCrea, J.D.; Mielke, E.W.; Ne'eman, Y. Progress in Metric Affine Gauge Theories of Gravity With Local Scale Invariance. *Found. Phys.* **1989**, *19*, 1075–1100. [CrossRef]
65. Hehl, F.W.; Socorro, J. Gauge theory of gravity: Electrically charged solutions within the metric affine framework. *Acta Phys. Polon. B* **1998**, *29*, 1113.
66. Ne'eman, Y.; Sijacki, D. Gravitational interaction of hadrons and leptons: Linear (multiplicity-free) bandor and nonlinear spinor unitary irreducible representations of SL(4,R). *Proc. Natl. Acad. Sci. USA* **1980**, *77*, 2353. [CrossRef] [PubMed]
67. Ne'eman, Y.; Sijacki, D. SL(4,R) World Spinors and Gravity. *Phys. Lett. B* **1985**, *157*, 275–279. [CrossRef]
68. Ne'eman, Y. World Spinors in Riemannian Gravity. DOE-ER40200-014. Available online: https://inis.iaea.org/search/search.aspx?orig_q=RN:17020101 (accessed on 13 November 2020).
69. Hehl, F.W.; McCrea, J.D.; Mielke, E.W.; Ne'eman, Y. Metric affine gauge theory of gravity: Field equations, Noether identities, world spinors, and breaking of dilation invariance. *Phys. Rept.* **1995**, *258*, 1–171. [CrossRef]
70. Sijacki, D. Affine particles and fields. *Int. J. Geom. Meth. Mod. Phys.* **2005**, *2*, 189–201. [CrossRef]
71. Hehl, F.W.; Macias, A. Metric affine gauge theory of gravity. 2. Exact solutions. *Int. J. Mod. Phys. D* **1999**, *8*, 399–416. [CrossRef]
72. Puetzfeld, D. Status of non-Riemannian cosmology. *New Astron. Rev.* **2005**, *49*, 59–64. [CrossRef]

73. Obukhov, Y.N.; Vlachynsky, E.J.; Esser, W.; Tresguerres, R.; Hehl, F.W. An Exact solution of the metric affine gauge theory with dilation, shear, and spin charges. *Phys. Lett. A* **1996**, *220*, 1–9. [CrossRef]
74. Rumpf, H. Quasiclassical limit of the Dirac equation and the equivalence principle in the Riemann-Cartan geometry. *NATO Sci. Ser. B* **1980**, *58*, 93–104.
75. Audretsch, J. Dirac Electron in Space-times With Torsion: Spinor Propagation, Spin Precession, and Nongeodesic Orbits. *Phys. Rev. D* **1981**, *24*, 1470–1477. [CrossRef]
76. Ni, W.T. Probing the microscopic origin of gravity via precision polarization and spin experiments. *Chin. Phys. Lett.* **2005**, *22*, 33.
77. Nitsch, J.; Hehl, F.W. Translational Gauge Theory of Gravity: Postnewtonian Approximation and Spin Precession. *Phys. Lett. B* **1980**, *90*, 98–102. [CrossRef]
78. Lammerzahl, C. Constraints on space-time torsion from Hughes-Drever experiments. *Phys. Lett. A* **1997**, *228*, 223–231. [CrossRef]
79. Cabral, F.; Lobo, F.S.N.; Rubiera-Garcia, D. Imprints from a Riemann-Cartan space-time on the energy levels of Dirac spinors. in press.
80. Puetzfeld, D.; Obukhov, Y.N. Probing non-Riemannian spacetime geometry. *Phys. Lett. A* **2008**, *372*, 6711–6716. [CrossRef]
81. Hayashi, K.; Shirafuji, T. New General Relativity. *Phys. Rev. D* **1979**, *19*, 3524–3553. [CrossRef]
82. Aldrovandi, R.; Pereira, J.G. *Teleparallel Gravity: An Introduction*; Fundamental Theories of Physics; Springer: Dordrecht, The Netherlands, 2013; Volume 173.
83. Maluf, J.W. The teleparallel equivalent of general relativity. *Ann. Phys.* **2013**, *525*, 339–357. [CrossRef]
84. Beltrán Jiménez, J.; Heisenberg, L.; Koivisto, T. Coincident General Relativity. *Phys. Rev. D* **2018**, *98*, 44048. [CrossRef]
85. Beltrán Jiménez, J.; Heisenberg, L.; Koivisto, T.S. Teleparallel Palatini theories. *JCAP* **2018**, *8*, 039. [CrossRef]
86. Järv, L.; Rünkla, M.; Saal, M.; Vilson, O. Nonmetricity formulation of general relativity and its scalar-tensor extension. *Phys. Rev. D* **2018**, *97*, 124025. [CrossRef]
87. Iosifidis, D.; Petkou, A.C.; Tsagas, C.G. Torsion/non-metricity duality in f(R) gravity. *Gen. Rel. Grav.* **2019**, *51*, 66. [CrossRef]
88. Harko, T.; Koivisto, T.S.; Lobo, F.S.N.; Olmo, G.J.; Rubiera-Garcia, D. Coupling matter in modified Q gravity. *Phys. Rev. D* **2018**, *98*, 084043. [CrossRef]
89. Obukhov, Y.N.; Hehl, F.W. Space-time metric from linear electrodynamics. *Phys. Lett. B* **1999**, *458*, 466–470. [CrossRef]
90. Barausse, E.; Berti, E.; Hertog, T.; Hughes, S.A.; Jetzer, P.; Pani, P.; Sotiriou, T.P.; Tamanini, N.; Witek, H.; Yagi, K.; et al. Prospects for Fundamental Physics with LISA. *Gen. Rel. Grav.* **2020**, *52*, 81. [CrossRef]
91. Lobo, F.S.N.; Olmo, G.J.; Rubiera-Garcia, D. Crystal clear lessons on the microstructure of spacetime and modified gravity. *Phys. Rev. D* **2015**, *91*, 124001. [CrossRef]
92. Harko, T.; Lobo, F.S.N. *Extensions of $f(R)$ Gravity: Curvature-Matter Couplings and Hybrid Metric-Palatini Theory*; Cambridge University Press: Cambridge, UK, 2018.
93. Berti, E.; Barausse, E.; Cardoso, V.; Gualtieri, L.; Pani, P.; Sperhake, U.; Stein, L.C.; Wex, N.; Yagi, K.; Baker, T.; et al. Testing General Relativity with Present and Future Astrophysical Observations. *Class. Quant. Grav.* **2015**, *32*, 243001. [CrossRef]
94. Nojiri, S.; Odintsov, S.D.; Oikonomou, V.K. Modified Gravity Theories on a Nutshell: Inflation, Bounce and Late-time Evolution. *Phys. Rept.* **2017**, *692*, 1–104. [CrossRef]

universe

MDPI

Review

Dirac, Majorana, Weyl in 4D

Loriano Bonora [1,2,*], Roberto Soldati [3] and Stav Zalel [4]

1 International School for Advanced Studies (SISSA), via Bonomea 265, 34136 Trieste, Italy
2 INFN-Sezione di Trieste, Padriciano, 99, 34149 Trieste TS, Italy
3 Dipartimento di Fisica e Astronomia, via Irnerio 46, 40126 Bologna, Italy; roberto.soldati@bo.infn.it
4 Blackett Laboratory, Imperial College, London SW7 2AZ, UK; stav.zalel11@imperial.ac.uk
* Correspondence: bonora@sissa.it

Received: 10 June 2020; Accepted: 30 July 2020; Published: 4 August 2020

Abstract: This is a review of some elementary properties of Dirac, Weyl and Majorana spinors in 4D. We focus in particular on the differences between massless Dirac and Majorana fermions, on one side, and Weyl fermions, on the other. We review in detail the definition of their effective actions, when coupled to (vector and axial) gauge fields, and revisit the corresponding anomalies using the Feynman diagram method with different regularisations. Among various well known results we stress in particular the regularisation independence in perturbative approaches, while not all the regularisations fit the non-perturbative ones. As for anomalies, we highlight in particular one perhaps not so well known feature: the rigid relation between chiral and trace anomalies.

Keywords: spinors in 4d; regularization; anomalies

1. Introduction

This paper is a review concerning the properties of Dirac, Weyl and Majorana fermions in a 4 dimensional Minkowski space-time. Fermions are unintuitive objects, thus the more fascinating. The relevant literature is enormous. Still problems that seem to be well understood, when carefully put under scrutiny, reveal sometimes unexpected aspects. The motivation for this paper is the observation that while Dirac fermions are very well known, both from the classical and the quantum points of view, Weyl and Majorana fermions are often treated as poor relatives [1] of the former, and, consequently, not sufficiently studied, especially for what concerns their quantum aspects. The truth is that these three types of fermions, while similar in certain respects, behave radically differently in others. Dirac spinors belong to a reducible representation of the Lorentz group, which can be irreducibly decomposed in two different ways: the first in eigenstates of the charge conjugation operator (Majorana), the second in eigenstates of the chirality operator (Weyl). Weyl spinors are bound to preserve chirality, therefore do not admit a mass term in the action and are strictly massless. Dirac and Majorana fermions can be massive. In this review we will focus mostly on massless fermions, and one of the issues we wish to elaborate on is the difference between massless Dirac, Majorana and Weyl fermions.

The key problem one immediately encounters is the construction of the effective action of these fermions coupled to gauge or gravity potentials. Formally, the effective action is the product of the eigenvalues of the relevant kinetic operator. The actual calculus can be carried out either perturbatively or non-perturbatively. In the first case the main approach is by Feynman diagrams, in the other case by analytical methods, variously called Seeley–Schwinger–DeWitt or heat kernel methods. While the procedure is rather straightforward in the Dirac (and Majorana) case, the same approach in the

1 Actually, the Weyl spinors are the main bricks of the original Standard Model, while the neutral Majorana spinors could probably be the basic particle bricks of Dark Matter, if any.

Weyl case is strictly speaking inaccessible. In this case, one has to resort to a roundabout method, the discussion of which is one of the relevant topics of this review. In order to clarify some basic concepts we carry out a few elementary Feynman diagram calculations with different regularisations (mostly Pauli–Villars and dimensional regularisation). The purpose is to justify the methods used to compute the Weyl effective action. A side bonus of this discussion is a clarification concerning the nonperturbative methods and the Pauli–Villars (PV) regularisation: contrary to the dimensional regularisation, the PV regularisation is unfit to be extended to the heat kernel-like methods, unless one is unwisely willing to violate locality.

A second major ground on which Weyl fermions split from Dirac and Majorana fermions is the issue of anomalies. To illustrate it in a complete and exhaustive way we limit ourselves here to fermion theories coupled to external gauge potentials and, using the Feynman diagrams, we compute all the anomalies (trace and gauge) in such a background. These anomalies have been calculated elsewhere in the literature in manifold ways and since a long time, so that there is nothing new in our procedure. Our goal here is to give a panoptic view of these computations and their interrelations. The result is interesting. Not only does one get a clear vantage point on the difference between Dirac and Weyl anomalies, but, for instance, it transpires that the rigid link between chiral and trace anomalies is not a characteristic of supersymmetric theories alone, but holds in general.

The paper is organized as follows. Section 2 is devoted to basic definitions and properties of Dirac, Weyl and Majorana fermions, in particular to the differences between massless Majorana and Weyl fermions. In Section 3, we discuss the problem related to the definition of a functional integral for Weyl fermions. In Section 4, we introduce perturbative regularisations for Weyl fermions coupled to vector potentials and verify that the addition of a free Weyl fermions of opposite handedness allows us to define a functional integral for the system, while preserving the Weyl fermion's chirality. Section 5 is devoted to an introduction to quantum Majorana fermions. In Section 6, we recalculate consistent and covariant gauge anomalies for Weyl and Dirac fermions, by means of the Feynman diagram technique. In particular, in Section 7, we do the same calculation in a vector-axial background, and in Section 8 we apply these results to the case of Majorana fermions. In Section 9, we compute also the trace anomalies of Weyl fermions due to the presence of background gauge potentials and show that they are rigidly related to the previously calculated gauge anomalies. Section 10 is devoted to a summary of the results. Three Appendices contain auxiliary material.

Historical references for this review are [1–16].

Notation

We use a metric $g_{\mu\nu}$ with mostly $(-)$ signature. The gamma matrices satisfy $\{\gamma^\mu, \gamma^\nu\} = 2g^{\mu\nu}$ and

$$\gamma_\mu^\dagger = \gamma_0 \gamma_\mu \gamma_0.$$

At times we use also the α matrices, defined by $\alpha_\mu = \gamma_0 \gamma_\mu$. The generators of the Lorentz group are $\Sigma_{\mu\nu} = \frac{1}{4}[\gamma_\mu, \gamma_\nu]$. The charge conjugation operator C is defined to satisfy

$$\gamma_\mu^T = -C^{-1}\gamma_\mu C, \qquad CC^* = -1, \qquad CC^\dagger = 1. \tag{1}$$

For example, $C = C^\dagger = C^{-1} = \gamma^0\gamma^2 = \alpha^2$ does satisfy all the above requirements, but it holds only in some γ-matrix representations, such as the Dirac and Weyl ones, not in the Majorana. The chiral matrix $\gamma_5 = i\gamma^0\gamma^1\gamma^2\gamma^3$ has the properties

$$\gamma_5^\dagger = \gamma_5, \qquad (\gamma_5)^2 = 1, \qquad C^{-1}\gamma_5 C = \gamma_5^T.$$

2. Dirac, Majorana and Weyl Fermions in 4D

Let us start from a few basic definitions and properties of spinors on a 4D Minkowski space [2]. A 4-component Dirac fermion ψ under a Lorentz transformation transforms as

$$\psi(x) \to \psi'(x') = \exp\left[-\frac{1}{2}\lambda^{\mu\nu}\Sigma_{\mu\nu}\right]\psi(x), \tag{2}$$

for $x'^{\mu} = \Lambda^{\mu}{}_{\nu}\, x^{\nu}$. Here $\lambda^{\mu\nu} + \lambda^{\nu\mu} = 0$ are six real canonical coordinates for the Lorentz group, $\Sigma_{\mu\nu}$ are the generators in the 4D reducible representation of Dirac bispinors, while $\Lambda^{\mu}{}_{\nu}$ are the Lorentz matrices in the irreducible vector representation $D(\frac{1}{2},\frac{1}{2})$. The invariant kinetic Lagrangian for a free Dirac field is

$$i\bar{\psi}\gamma^{\mu}\partial_{\mu}\psi. \tag{3}$$

where $\bar{\psi} = \psi^{\dagger}\gamma_0$.

A Dirac fermion admits a Lorentz invariant mass term $m\bar{\psi}\psi$.

A Dirac bispinor can be seen as the direct sum of two Weyl spinors

$$\psi_L = P_L\psi, \qquad \psi_R = P_R\psi, \qquad \text{where} \qquad P_L = \frac{1-\gamma_5}{2}, \qquad P_R = \frac{1+\gamma_5}{2}$$

with opposite chiralities

$$\gamma_5\psi_L = -\psi_L, \qquad \gamma_5\psi_R = \psi_R.$$

A left-handed Weyl fermion admits a Lagrangian kinetic term

$$i(\psi_L, \gamma\cdot\partial\psi_L) = i\overline{\psi}_L\gamma^{\mu}\partial_{\mu}\psi_L \tag{4}$$

but not a mass term, because $(\psi_L, \psi_L) = 0$, since $\gamma_5\gamma^0 + \gamma^0\gamma_5 = 0$. A Weyl fermion is massless and this property is protected by the chirality conservation.

For Majorana fermions we need the notion of *Lorentz covariant conjugate* spinor, $\hat{\psi}$:

$$\hat{\psi} = \gamma_0 C\psi^*. \tag{5}$$

It is not hard to show that if ψ transforms like (2), then

$$\hat{\psi}(x) \to \hat{\psi}'(x') = \exp\left[-\frac{1}{2}\lambda^{\mu\nu}\Sigma_{\mu\nu}\right]\hat{\psi}(x). \tag{6}$$

Therefore one can impose on ψ the condition

$$\psi = \hat{\psi} \tag{7}$$

because both sides transform in the same way. By definition, a spinor satisfying (7) is a Majorana spinor. It admits both kinetic and mass term.

It is a renowned fact that the group theoretical approach [1] to Atiyah Theory is one of the most solid and firm pillars in modern Physics. To this concern, the contributions by Eugene Paul Wigner were of invaluable importance [3]. In terms of Lorentz group representations we can say the following. γ_5 commutes with Lorentz transformations $\exp\left[-\frac{1}{2}\lambda^{\mu\nu}\Sigma_{\mu\nu}\right]$. So do P_L and P_R. This means that the Dirac

[2] This and the following section are mostly based on [17,18].

representation is reducible. Multiplying the spinors by P_L and P_R selects irreducible representations, the Weyl ones. To state it more precisely, Weyl representations are irreducible representations of the group $SL(2,C)$, which is the covering group of the proper ortochronous Lorentz group. They are usually denoted $(\frac{1}{2},0)$ and $(0,\frac{1}{2})$ in the $SU(2)\times SU(2)$ notation of the $SL(2,C)$ irreps. As we have seen in (6), Lorentz transformations commute also with the charge conjugation operation

$$\mathcal{C}\psi\mathcal{C}^{-1} = \eta_C\gamma_0 C\psi^* \tag{8}$$

where η_C is a phase which, for simplicity, in the sequel we set equal to 1. This also implies that Dirac spinors are reducible and suggests another possible reduction: by imposing (7) we single out another irreducible representation, the Majorana one. The Majorana representation is the minimal irreducible representation of a (one out of eight) covering of the complete Lorentz group [5,11]. It is evident, and well-known, that Majorana and Weyl representations in 4D are incompatible.

Let us consider next the charge conjugation and parity and recall the relevant properties of a Weyl fermion. We have

$$\mathcal{C}\psi_L\mathcal{C}^{-1} = P_L\mathcal{C}\psi\mathcal{C}^{-1} = P_L\hat{\psi} = \hat{\psi}_L. \tag{9}$$

The charge conjugate of a Majorana field is, by definition, itself. While the action of a Majorana field is invariant under charge conjugation, for a Weyl fermion we have

$$\mathcal{C}\left(\int i\overline{\psi_L}\gamma^\mu\partial_\mu\psi_L\right)\mathcal{C}^{-1} = \int i\overline{\hat{\psi}_L}\gamma^\mu\partial_\mu\hat{\psi}_L = \int i\overline{\psi_R}\gamma^\mu\partial_\mu\psi_R. \tag{10}$$

i.e., a Weyl fermion is, so to say, maximally non-invariant.

The parity operation is defined by

$$\mathcal{P}\psi_L(t,\vec{x})\mathcal{P}^{-1} = \eta_P\gamma_0\psi_R(t,-\vec{x}) \tag{11}$$

where η_P is a phase, which in the sequel we set to 1. In terms of the action we have

$$\mathcal{P}\left(\int \overline{\psi}_L\gamma^\mu\partial_\mu\psi_L\right)\mathcal{P}^{-1} = \int \overline{\psi}_R\gamma^\mu\partial_\mu\psi_R, \tag{12}$$

For a Majorana fermion the action is invariant under parity.

This also suggests a useful representation for a Majorana fermion. Let $\psi_R = P_R\psi$ be a generic Weyl fermion. We have $P_R\psi_R = \psi_R$ and it is easy to prove that $P_L\widehat{\psi_R} = \widehat{\psi_R}$, i.e., $\widehat{\psi_R}$ is left-handed. Therefore the sum $\psi_M = \psi_R + \widehat{\psi_R}$ is a Majorana fermion because it satisfies (7). Any Majorana fermion can be represented in this way. This representation is instrumental in the calculus of anomalies, see below.

Considering next CP, from the above it follows that the action of a Majorana fermion is obviously invariant under it. On the other hand, for a Weyl fermion we have

$$\mathcal{CP}\psi_L(t,\vec{x})(\mathcal{CP})^{-1} = \gamma_0 P_R\hat{\psi}(t,-\vec{x}) = \gamma_0\hat{\psi}_R(t,-\vec{x}). \tag{13}$$

Applying, now, CP to the action for a Weyl fermion, one gets

$$\mathcal{CP}\left(\int i\overline{\psi_L}\gamma^\mu\partial_\mu\psi_L\right)(\mathcal{CP})^{-1} = \int i\overline{\hat{\psi}_R}(t,-\vec{x})\gamma^{\mu+}\partial_\mu\hat{\psi}_R(t,-\vec{x}) = \int i\overline{\hat{\psi}_R}(t,\vec{x})\gamma^\mu\partial_\mu\hat{\psi}_R(t,\vec{x}). \tag{14}$$

One can prove as well that

$$\int i\overline{\hat{\psi}_R}(t,\vec{x})\gamma^\mu\partial_\mu\hat{\psi}_R(t,\vec{x}) = \int i\overline{\psi_L}(x)\gamma^\mu\partial_\mu\psi_L(x). \tag{15}$$

Therefore the action for a Weyl fermion is CP invariant. It is also, separately, T invariant, and, so, CPT invariant. The transformation properties of the Weyl and Majorana spinor fields are summarized in the following table:

$$
\begin{array}{lcc}
 & \text{Majorana} & \text{Weyl} \\
\text{P}: & \mathcal{P}\psi(t,\vec{x})\mathcal{P}^{-1} = \gamma_0\psi(t,-\vec{x}) & \mathcal{P}\psi_L(t,\vec{x})\mathcal{P}^{-1} = \gamma_0\psi_R(t,-\vec{x}) \\
\text{C}: & \mathcal{C}\psi\mathcal{C}^{-1} = \gamma_0 C\psi^* = \psi & \mathcal{C}\psi_L\mathcal{C}^{-1} = P_L\hat{\psi} = \hat{\psi}_L \\
\text{CP}: & \mathcal{C}\mathcal{P}\psi(t,\vec{x})(\mathcal{C}\mathcal{P})^{-1} = \psi(t,-\vec{x}) & \mathcal{C}\mathcal{P}\psi_L(t,\vec{x})(\mathcal{C}\mathcal{P})^{-1} = \gamma_0\hat{\psi}_R(t,-\vec{x})
\end{array}
\tag{16}
$$

The quantum interpretation of the field ψ_L starts from the plane wave expansion

$$\psi_L(x) = \int dp\, \left(a(p)u_-(p)e^{-ipx} + b^\dagger(p)v_+(p)e^{ipx}\right) \tag{17}$$

where u_-, v_+ are fixed and independent left-handed spinors (there are only two of them). Such spin states are explicitly constructed in Appendix A. We interpret (17) as follows: $b^\dagger(p)$ creates a left-handed particle while $a(p)$ destroys a left-handed particle with negative helicity (because of the opposite momentum). However, Equations (13) and (14) force us to identify the latter with a right-handed antiparticle: C maps particles to antiparticles, while P invert helicities, so CP maps left-handed particles to right-handed antiparticles. One need not stress that in this game right-handed particles or left-handed antiparticles are absent.

Remark 1. *Let us comment on a few deceptive possibilities for a mass term for a Weyl fermion. A mass term $\bar{\psi}\psi$ for a Dirac spinor can also be rewritten by projecting the latter into its chiral components* [3]

$$\bar{\psi}\psi = \overline{\psi_L}\psi_R + \overline{\psi_R}\psi_L. \tag{18}$$

If ψ is a Majorana spinor, $\psi = \hat{\psi}$, this can be rewritten as

$$\bar{\hat{\psi}}\hat{\psi} = \overline{\psi_L}\hat{\psi}_R + \overline{\hat{\psi}_R}\psi_L, \tag{19}$$

which is, by construction, well defined and Lorentz invariant. Now, by means of the Lorentz covariant conjugate, we can rewrite (19) as

$$(\psi_L)^T C^{-1}\psi_L + \psi_L^\dagger C(\psi_L)^*, \tag{20}$$

which is expressed only in terms of ψ_L, although contains both chiralities. (20) may induce the impression that there exists a mass term also for Weyl fermions. This is not so. If we add this term to the kinetic term (4), the ensuing equation of motion is not Lorentz covariant: the kinetic and mass term in it belong to two different representations. To be more explicit, a massive Dirac equation of motion for a Weyl fermion should be

$$i\gamma^\mu\partial_\mu\psi_L - m\psi_L = 0, \tag{21}$$

but it breaks Lorentz covariance because the first piece transforms according to the $(0,\frac{1}{2})$ representation, while the second according to $(\frac{1}{2},0)$, and is not Lagrangian [4]*. The reason is, of course, that (20) is not expressible in*

3 Here and in the sequel we use $\bar{\psi}$ and $\overline{\psi}$, $\hat{\psi}$ and $\widehat{\psi}$ in a precise sense, which is hopefully unambiguous. For instance $\bar{\psi} = \psi^\dagger\gamma_0$, $\overline{\psi_R} = (\psi_R)^\dagger\gamma_0$, $\widehat{\psi_R} = \gamma_0 C(\psi_R)^*$, $\hat{\psi}_R = P_R\hat{\psi}$ and so on.

4 Instead of the second term in the LHS of (21) one could use $mC\overline{\psi_L}^T$, which has the right Lorentz properties, but the corresponding Lagrangian term would not be self-adjoint and one would be forced to introduce the adjoint term and end up again with (20). This implies, in particular, that there does not exist such a thing as a "massive Weyl propagator", that is a massive propagator involving only one chirality.

the same canonical form as (4). This structure is clearly visible in the four component formalism used so far, but much less recognizable in the two-component formalism.

Weyl Fermions and Massless Majorana Fermions

That a massive Majorana fermion and a Weyl fermion are different objects is uncontroversial. The question whether a massless Majorana fermion is or is not the same as a Weyl fermion at both the classical level and the quantum level is, instead, not always clear in the literature. Let us consider the simplest case in which there is no quantum number appended to the fermion. To start let us recall the obvious differences between the two. The first, and most obvious, has already been mentioned: they belong to two different irreducible representations of the Lorentz group (in 4D there cannot exist a spinor that is simultaneously Majorana and Weyl, like in 2D and 10D). Another important difference is that the helicity for a Weyl fermion is well defined and corresponds to its chirality, while for a Majorana fermion chirality is undefined, so that the relation with its helicity is also undefined. Then, a parity operation maps the Majorana action into itself, while it maps the Weyl action (4) into the same action for the opposite chirality. The same holds for the charge conjugation operator. Why they are sometimes considered the same object may be due to the fact that we can establish a one-to-one correspondence between the components of a Weyl spinor and those of a Majorana spinor in such a way that the Lagrangian, in two-component notation, looks the same, see, for instance [16]. If, in the chiral representation, we write ψ_L as $\begin{pmatrix} \omega \\ 0 \end{pmatrix}$, where ω is a two component spinor, then (4) above becomes

$$i\omega^\dagger \bar{\sigma}^\mu \partial_\mu \omega \tag{22}$$

which has the same form (up to an overall factor) as a massless Majorana action (see Section 5 below and Equation (49)). The action is not everything in a theory, it must be accompanied by a set of specifications. Even though numerically the actions coincide, the way they respond to a variation of the Weyl and Majorana fields is different. One leads to the Weyl equation of motion, the other to the Majorana equation of motion. The delicate issue is precisely this: when we take the variation of an action with respect to a field in order to extract the equations of motion, we must make sure that the variation respects the symmetries and the properties that are expected in the equations of motion [5]. Specifically in the present case, if we wish the equation of motion to preserve chirality we must use variations that preserve chirality, i.e., variations that are eigenfunctions of γ_5. If instead we wish the equation of motion to transform in the Majorana representation we have to use variations that transform suitably, i.e., variations that are eigenfunctions of the charge conjugation operator. If we do so, we obtain two different results which are irreducible to each other no matter which action we use.

There is no room for confusing massless Majorana spinors with chiral Weyl spinors. A classical Majorana spinor is a self-conjugated bispinor, that can always be chosen to be real and always contains both chiralities in terms of four real independent component functions. It describes neutral spin 1/2 objects—not yet detected in Nature—and consequently there is no phase transformation (U(1) continuous symmetry) involving self-conjugated Majorana spinors, independently of the presence or not of a mass term. Hence, e.g., its particle states do not admit antiparticles of opposite charge, simply because charge does not exist at all for charge self-conjugated spinors (actually, this was the surprising discovery of Ettore Majorana, after the appearance of the Dirac equation and the positron detection). The general solution of the wave field equations for a free Majorana spinor always entails the presence of two polarization states with opposite helicity. On the contrary, it is well known that a chiral Weyl spinor, describing massless neutrinos in the Standard Model, admits only one polarization or helicity

---

5 For instance, in gravity theories, the metric variation $\delta g_{\mu\nu}$ is generic while not ceasing to be a symmetric tensor.

state, it always involves antiparticles of opposite helicity and it always carries a conserved internal quantum number such as the lepton number, which is opposite for particles and antiparticles.

Finally, and most important, in the quantum theory a crucial role is played by the functional measure, which is different for Weyl and Majorana fermions. We will shortly come back to this point. Before that, it is useful to clarify an issue concerning the just mentioned U(1) continuous symmetry of Weyl fermions. The latter is sometime confused with the axial $\mathbb{R}$ symmetry of Majorana fermions and assumed to justify the identification of Weyl and massless Majorana fermions. To start with, let us consider a free massless Dirac fermion ψ. Its free action is clearly invariant under the transformation $\delta\psi = i(\alpha + \gamma_5\beta)\psi$, where α and β are real numbers. This symmetry can be gauged by minimally coupling ψ to a vector potential V_μ and an axial potential A_μ, in the combination $V_\mu + \gamma_5 A_\mu$, so that α and β become arbitrary real functions. For convenience, let us choose the Majorana representation for gamma matrices, so that all of them, including γ_5, are imaginary. If we now impose ψ to be a Majorana fermion, its four component can be chosen to be real and only the symmetry parametrized by β makes sense in the action (let us call it β symmetry). If instead we impose ψ to be Weyl, say $\psi = \psi_L$, then , since $\gamma_5\psi_L = \psi_L$, the symmetry transformation will be $\delta\psi_L = i(\alpha - \beta)\psi_L$.

We believe this may be the origin of the confusion, because it looks like we can merge the two parameters α and β into a single parameter identified with the β of the Majorana axial β symmetry. However this is not correct because for a right handed Weyl fermion the symmetry transformation is $\delta\psi_R = i(\alpha + \beta)\psi_R$. Forgetting β, the Majorana fermion does not transform. Forgetting α, both Weyl and Majorana fermions transform, but the Weyl fermions transform with opposite signs for opposite chiralities. This distinction will become crucial in the computation of anomalies (see below).

3. Functional Integral for Dirac, Weyl and Majorana Fermions

In quantum field theory there is one more reason to distinguish between massless Weyl and Majorana fermions: their functional integration measure is formally and substantially different. Although the action in the two-component formalism may take the same form (22) for both, the change of integration variable from ψ_L to ω is not an innocent field redefinition because the functional integration measure changes. The purpose of this section is to illustrate this issue. To start with, let us clarify that speaking about functional integral measure is a colourful but not rigorous parlance. The real issue here is the definition of the functional determinant for a Dirac-type matrix-valued differential operator.

Let us start with some notations and basic facts. We denote by $\not{D}$ the standard Dirac operator: namely, the massless matrix-valued differential operator applied in general to Dirac spinors on the 4D curved space with Minkowski signature $(+,-,-,-)$

$$\not{D} = i\,(\not{\partial} + \not{V}) \tag{23}$$

where V_μ is any anti-Hermitean vector potential, including a spin connection in the presence of a non-trivial background metric. We use here the four component formalism for fermions. The functional integral, i.e., the effective action for a quantum Dirac spinor in the presence of a classical background potential

$$\mathcal{Z}[V] = \int \mathcal{D}\psi\mathcal{D}\bar{\psi}\; e^{i\int d^4x\,\sqrt{g}\,\bar{\psi}\not{D}\psi} \tag{24}$$

is formally understood as the determinant of $\not{D}$: $\det(\not{D}) = \prod_i^\infty \lambda_i$. From a concrete point of view, the latter can be operatively defined in two alternative ways: either in perturbation theory, i.e., as the sum of an infinite number of 1-loop Feynman diagrams, some of which contain UV divergences by naïve power counting, or by a non-perturbative approach, i.e., as the suitably regularised infinite product of the eigenvalues of $\not{D}$ by means of the analytic continuation tool. It is worthwhile to remark that, on the one hand, the perturbative approach requires some UV regulator and renormalisation prescription, in order to give a meaning to a finite number of UV divergent 1-loop diagrams by naïve power counting. On the the other hand, in the non-perturbative framework the complex power construction and the analytic continuation tool, if available, provide by themselves the whole necessary setting up to define the infinite product of the eigenvalues of a normal operator, without need of any further regulator.

In many practical calculations one has to take variations of (24) with respect to V. In turn, any such variation requires the existence of an inverse of the kinetic operator, as follows from the abstract formula for the determinant of an operator A

$$\delta \mathrm{detA} = \mathrm{detA}\,\mathrm{tr}\left(\mathrm{A}^{-1}\delta\mathrm{A}\right).$$

It turns out that an inverse of $\not{D}$ does exist and, if full causality is required in forwards and backwards time evolution on e.g., Minkowski space, it is the Feynman propagator or Schwinger distribution $\not{S}$, which is unique and characterized by the well-known Feynman prescription, in such a manner that

$$\not{D}_x \not{S}(x-y) = \delta(x-y), \qquad \not{D}\not{S} = 1. \tag{25}$$

The latter is a shortcut operator notation, which we are often going to use in the sequel [6].

For instance, the scheme to extract the trace of the stress-energy tensor from the functional integral is well-known. It is its response under a Weyl (or even a scale) transform $\delta_\omega g_{\mu\nu} = 2\omega g_{\mu\nu}$:

$$\delta_\omega \log \mathcal{Z} = \int d^4x\,\omega(x)\,g_{\mu\nu}(x)\,\langle\,T^{\mu\nu}(x)\rangle \tag{26}$$

where $g_{\mu\nu}(x)\langle T^{\mu\nu}(x)\rangle$ is the quantum trace of the energy-momentum tensor. Analogously, the divergence of the vector current $j_\mu = \bar{\psi}\gamma_\mu\psi$ is the response of $\log\mathcal{Z}$ under the Abelian gauge transformation $\delta_\lambda V_\mu = \partial_\mu\lambda$:

$$\delta_\lambda \log \mathcal{Z} = -i\int d^4x\,\lambda(x)\partial_\mu\langle j^\mu(x)\rangle \tag{27}$$

and so on. These quantities can be calculated in various ways with perturbative or non-perturbative methods. The most frequently used ones are the Feynman diagram technique and the so-called analytic functional method, respectively. The latter denomination actually includes a collection of approaches, ranging from the Schwinger's proper-time method [6] to the heat kernel method [13], the Seeley-DeWitt [7,8] and the zeta-function regularisation [10]. The central tool in these approaches is the (full) kinetic operator of the fermion action (or the square thereof), and its inverse, the full fermion propagator. All these methods yield well-known results with no disagreement among them.

On the contrary, when one comes to Weyl fermions things drastically change. The classical action on the 4D Minkowski space for a left-handed Weyl fermion reads

$$S_L = \int \mathrm{d}^4x\,\bar{\psi}_L \not{D}\psi_L. \tag{28}$$

The Dirac operator, acting on left-handed spinors maps them to right-handed ones. Hence, the Sturm–Liouville or eigenvalue problem itself is not well posed, so that the Weyl determinant cannot even be defined. This is reflected in the fact that the inverse of $\not{D}_L = \not{D}P_L = P_R\not{D}$ does not exist, since it is the product of an invertible operator times a projector. As a consequence the full propagator

6 For simplicity we understand factors of $\sqrt{g}$, which should be there, see [7], but are inessential in this discussion.

of a Weyl fermion does not exist in this naïve form (this problem can be circumvented in a more sophisticated approach, see below) [7].

The lack of an inverse for the chiral Dirac–Weyl kinetic term has drastic consequences even at the free non-interacting level. For instance, the evaluation of the functional integral (i.e., formally integrating out the spinor fields) involves the inverse of the kinetic operator: thus, it is clear that the corresponding formulas for the chiral Weyl quantum theory cannot exist at all, so that no Weyl effective action can be actually defined in this way even in the free non-interacting case. Let us add that considering the square of the kinetic operator, as it is often done in the literature, does not change this conclusion.

It may sound strange that the (naïve) full propagator for Weyl fermions does not exist, especially if one has in mind perturbation theory in Minkowski space. In that case, in order to construct Feynman diagrams, one uses the ordinary free Feynman propagator for Dirac fermions. The reason one can do so is because the information about chirality is preserved by the fermion–boson–fermion vertex, which contains the P_L projector (the use of a free Dirac propagator is formally justified, because one can add a free right-handed fermion to allow the inversion of the kinetic operator, see below). On the contrary, the full (non-perturbative) propagator is supposed to contain the full chiral information, including the information contained in the vertex, i.e., the potential, as it will be explicitly checked here below. In this problem there is no simple shortcut such as pretending to replace the full Weyl propagator with the full Dirac propagator multiplied by a chiral projector, because this would destroy any information concerning the chirality.

The remedy for the Weyl fermion disaster is to use as kinetic operator

$$i\gamma^{\mu}\left(\partial_{\mu}+P_{L}V_{\mu}\right), \tag{30}$$

which is invertible and in accord with the above mentioned Feynman diagram approach. It corresponds to the intuition that the free right-handed fermions added to the left-handed theory in this way do not interfere with the conservation of chirality and do not alter the left-handed nature of the theory. It is important to explicitly check it. The next section is devoted to a close inspection of this problem and its solution.

4. Regularisations for Weyl Spinors

The classical Lagrange density for a Weyl (left) spinor in the four component formalism

$$\psi(x)=\chi_{L}(x)=\begin{pmatrix}\chi(x)\\0\end{pmatrix}$$

reads

$$\mathcal{K}(x)=\overline{\psi}(x)\,i\partial\!\!\!/\,\psi(x)=\chi_{L}^{\dagger}(x)\,\alpha^{\nu}i\partial_{\nu}\chi_{L}(x). \tag{31}$$

It follows that the corresponding matrix valued Weyl differential operator

$$w_{L}\equiv\alpha^{\nu}i\partial_{\nu}P_{L} \tag{32}$$

[7] It is incorrect to pretend that the propagator is $\$_L = \$P_R = P_L\$$. First because such an inverse does not exist, second because, even formally,

$$D\!\!\!\!/_{L}\$_{L}=P_{R},\qquad\text{and}\qquad \$_{L}\overleftarrow{D\!\!\!\!/}_{L}=P_{L} \tag{29}$$

The inverse of the Weyl kinetic operator is not the inverse of the Dirac operator multiplied by a chiral projector. Therefore the propagator for a Weyl fermion is not the Feynman propagator for a Dirac fermion multiplied by the same projector.

is singular and does not possess any rank-four inverse. After minimal coupling with a real massless vector field $A^{\mu}(x)$ we come to the classical Lagrangian

$$\mathcal{L} = \chi_L^{\dagger}\,\alpha^{\nu} i\partial_{\nu}\chi_L + gA^{\nu}\,\chi_L^{\dagger}\,\alpha_{\nu}\chi_L - \tfrac{1}{4}\,F^{\mu\nu}\,F_{\mu\nu} \tag{33}$$

where $F_{\mu\nu} = \partial_{\mu}A_{\nu} - \partial_{\nu}A_{\mu}$. It turns out that the classical action

$$S = \int \mathrm{d}^4x\,\mathcal{L} \tag{34}$$

is invariant under the Poincaré group, as well as under the internal U(1) phase transformations $\chi_L(x) \;\mapsto\; e^{ig\theta}\chi_L(x)$. The action integral is invariant under the so called scale or dilatation transformations, viz.,

$$x'^{\mu} = e^{-\varrho}x^{\mu} \qquad \chi_L'(x) = e^{\frac{3}{2}\varrho}\chi_L(e^{\varrho}x) \qquad A'^{\mu}(x) = e^{\varrho}A^{\mu}(e^{\varrho}x)$$

with $\varrho \in \mathbb{R}$, as well as with respect to the local phase or gauge transformations

$$\chi_L'(x) = e^{ig\theta(x)}\chi_L(x) \qquad\qquad A_{\nu}'(x) = A_{\nu}(x) + \partial_{\nu}\theta(x)$$

which amounts to the ordinary U(1) phase transform in the limit of constant phase. It follows therefrom that there are twelve conserved charges in this model at the classical level and, in particular, owing to scale and gauge invariance, no mass term is allowed for both spinor and vector fields. The question naturally arises if those symmetries hold true after the transition to the quantum theory and, in particular, if they are protected against loop radiative corrections within the perturbative approach. Now, as explained above, in order to develop perturbation theory, one faces the problem of the lack of an inverse for both the Weyl and gauge fields, owing to chirality and gauge invariance. In order to solve it, it is expedient to add to the Lagrangian non-interacting terms, which are fully decoupled from any physical quantity. They break chirality and gauge invariance, albeit in a harmless way, just to allow us to define a Feynman propagator, or causal Green's functions, for both the Weyl and gauge quantum fields. The simplest choice, which preserves Poincaré and internal U(1) phase change symmetries, is provided by

$$\mathcal{L}' = \varphi_R^{\dagger}\,\alpha^{\nu} i\partial_{\nu}\varphi_R - \tfrac{1}{2}(\partial\cdot A)^2$$

where

$$\psi(x) = \varphi_R(x) = \begin{pmatrix} 0 \\ \varphi(x) \end{pmatrix}$$

is a left-chirality breaking right-handed Weyl spinor field. Notice *en passant* that the modified Lagrangian $\mathcal{L} + \mathcal{L}'$ exhibits a further U(1) internal symmetry under the so called chiral phase transformations

$$\psi'(x) = (\cos\theta + i\sin\theta\,\gamma_5)\psi(x) \qquad\qquad \psi(x) = \begin{pmatrix} \chi(x) \\ \varphi(x) \end{pmatrix}$$

so that the modified theory involves another conserved charge at the classical level. From the modified Lagrange density we get the Feynman propagators for the massless Dirac field $\psi(x)$, as well as for the massless vector field in the so called Feynman gauge: namely,

$$S(p) = \frac{i\not{p}}{p^2 + i\varepsilon} \qquad\qquad D_{\mu\nu}(k) = \frac{-i\eta_{\mu\nu}}{k^2 + i\varepsilon} \tag{35}$$

and the vertex $ig\gamma^\nu P_L$, with $k+p-q=0$, which involves a vector particle of momentum k and a Weyl pair of particle and anti-particle of momenta p and q, respectively, and of opposite helicity. [8]

Our purpose hereafter is to show that, notwithstanding the use of the non-chiral propagators (35), a mass in the Weyl kinetic term cannot arise as a consequence of quantum corrections. The lowest order 1-loop correction to the kinetic term $\not{k}P_L$ is provided by the Feynman rules in Minkowski space, in the following form

$$\Sigma_2(\not{k}) = -ig^2 \int \frac{\mathrm{d}^4\ell}{(2\pi)^4}\, \gamma^\mu D_{\mu\nu}(k-\ell)\, S(\ell)\, \gamma^\nu P_L. \tag{36}$$

A mass term in this context should be proportional to the identity matrix (in the spinor space).

By naïve power counting the above 1-loop integral turns out to be UV divergent. Hence, a regularisation procedure is mandatory to give a meaning and evaluate the radiative correction $\Sigma_2(\not{k})$ to the Weyl kinetic operator. Here in the sequel we shall examine in detail the dimensional, Pauli–Villars and UV cut-off regularisations.

4.1. Dimensional, PV and Cutoff Regularisations

In a 2ω-dimensional space-time, the dimensionally regularised radiative correction to the Weyl kinetic term takes the form

$$\mathtt{reg}\,\Sigma_2(\not{k}) = -ig^2\mu^{2\epsilon} \int \frac{\mathrm{d}^{2\omega}\ell}{(2\pi)^{2\omega}}\, D_{\mu\nu}(\ell)\, \gamma^\mu S(\ell+k)\, \gamma^\nu P_L \tag{37}$$

where $\epsilon = 2-\omega > 0$ is the shift with respect to the physical space-time dimensions. Since the above expression is traceless and has the canonical engineering dimension of a mass in natural units, it is quite apparent that the latter cannot generate any mass term, which, as anticipated above, would be proportional to the unit matrix. Hence, mass is forbidden and it remains for us to evaluate

$$\mathtt{reg}\,\Sigma_2(\not{k}) \equiv f(k^2)\,\not{k}P_L \qquad \mathrm{tr}\,[\not{k}\,\mathtt{reg}\,\Sigma_2(\not{k})] = \tfrac{1}{2}2^\omega k^2 f(k^2) \tag{38}$$

$$\mathrm{tr}\,[\not{k}\,\mathtt{reg}\,\Sigma_2(\not{k})] = g^2\mu^{2\epsilon}(2\pi)^{-2\omega}\int \mathrm{d}^{2\omega}\ell\, \frac{(-i)\,\mathrm{tr}\,(\not{k}\gamma^\lambda\not{\ell}\gamma_\lambda P_L)}{[(\ell-k)^2+i\varepsilon]\,(\ell^2+i\varepsilon)}. \tag{39}$$

For $2^\omega \times 2^\omega$ γ-matrix traces in a 2ω-dimensional space-time with a Minkowski signature the following formulas are necessary

$$\mathrm{tr}\,(\gamma^\mu\gamma^\nu) = g^{\mu\nu}\,\mathrm{tr}\,\mathbb{I} = 2^\omega g^{\mu\nu} \tag{40}$$

$$2^{-\omega}\mathrm{tr}\left(\gamma^\kappa\gamma^\lambda\gamma^\mu\gamma^\nu\right) = g^{\kappa\lambda}g^{\mu\nu} - g^{\kappa\mu}g^{\lambda\nu} + g^{\kappa\nu}g^{\lambda\mu}. \tag{41}$$

Then we get $\mathrm{tr}\,(\not{k}\gamma^\lambda\not{\ell}\gamma_\lambda P_L) = 2^\omega(\epsilon-1)\,p\cdot\ell$ and thereby

$$k^2 f(k^2) = ig^2\mu^{2\epsilon}\,\frac{\epsilon-1}{(2\pi)^{2\omega}}\int \frac{2p\cdot\ell\,\mathrm{d}^{2\omega}\ell}{[(\ell-k)^2+i\varepsilon]\,(\ell^2+i\varepsilon)}. \tag{42}$$

Turning to the Feynman parametric representation we obtain

$$k^2 f(k^2) = ig^2\mu^{2\epsilon}\,\frac{\epsilon-1}{(2\pi)^{2\omega}}\int_0^1 \mathrm{d}x \int \frac{2p\cdot\ell\,\mathrm{d}^{2\omega}\ell}{[\ell^2-2x\,k\cdot\ell+xk^2+i\varepsilon]^2}. \tag{43}$$

---

[8] Customarily, the on-shell 1-particle states of a left Weyl spinor field are a left-handed particle with negative helicity $-\frac{1}{2}\hbar$ and a right-handed antiparticle of positive helicity $\frac{1}{2}\hbar$.

Completing the square in the denominator and after shifting the momentum $\ell' \equiv \ell - xp$, dropping the linear term in ℓ' in the numerator owing to symmetric integration, we have

$$f(k^2) = 2ig^2\mu^{2\epsilon}\,\frac{\epsilon-1}{(2\pi)^{2\omega}}\int_0^1 \mathrm{d}x\,x\int\frac{\mathrm{d}^{2\omega}\ell}{[\,\ell^2+x(1-x)k^2+i\varepsilon\,]^2}. \tag{44}$$

One can perform the Wick rotation and readily get the result

$$\begin{aligned} f(k^2) &= -2g^2\mu^{2\epsilon}\,\frac{\epsilon-1}{(4\pi)^{\omega}}\int_0^1 \mathrm{d}x\,x\int_0^\infty \mathrm{d}\tau\,\tau^{\epsilon-1}\,e^{-\tau x(1-x)\,k_E^2} \\ &= 2\left(\frac{g}{4\pi}\right)^2[\,\Gamma(\epsilon)-\Gamma(1+\epsilon)\,]\left(-\frac{4\pi\mu^2}{k^2}\right)^{\epsilon} B(2-\epsilon,1-\epsilon). \end{aligned} \tag{45}$$

Expansion around $\epsilon = 0$ yields

$$f(k^2) = \left(\frac{g}{4\pi}\right)^2\left[\frac{1}{\epsilon}+1+3\mathbf{C}+\ln\left(-\frac{4\pi\mu^2}{k^2}\right)\right] + \text{evanescent} \tag{46}$$

where $\mathbf{C}$ denotes the Euler–Mascheroni constant.

Similar results are obtained with the Pauli–Villars and cut-off regularisations. In the PV case the latter is simply implemented by the following replacement of the massless Dirac propagator

$$\mathtt{reg}\,\Sigma_2(\not{k}) = -\,ig^2\int\frac{\mathrm{d}^4\ell}{(2\pi)^4}\,\gamma^\mu\,D_{\mu\nu}(\,k-\ell\,)\,\sum_{s=0}^{S} C_s\,S(\ell,\,M_s)\,\gamma^\nu P_L, \tag{47}$$

where $M_0 = 0,\ C_0 = 1$ while $\{\,M_s \equiv \lambda_s\,M\,|\,\lambda_s \gg 1\ (\,s = 1,2,\ldots,S\,)\,\}$ is a collection of very large auxiliary masses. The constants C_s are required to satisfy:

$$\sum_{s=1}^{S} C_s = -1 \qquad \sum_{s=1}^{S} C_s\lambda_s = 0$$

and the following identification with the divergent parameter is made

$$\frac{1}{\epsilon} = \sum_{s=1}^{S} C_s\,\ln\lambda_s.$$

The result for $f(k^2)$ is

$$f(k^2) = \left(\frac{g}{4\pi}\right)^2\left[\sum_{s=1}^{S} C_s\,\ln\lambda_s+\frac{1}{4}+\frac{1}{2}\ln\left(-\frac{M^2}{k^2}\right)\right] + \text{evanescent}. \tag{48}$$

The same calculation can be repeated with an UV cutoff K, see [18]. To sum up, we have verified that the 1-loop correction to the (left) Weyl spinor self-energy has the general form, which is universal, i.e., regularisation independent: namely,

$$\begin{aligned} \mathtt{reg}\,\Sigma_2(\not{k}) &\equiv f(k^2)\,\not{k}P_L \\ f(k^2) &:= \left(\frac{g}{4\pi}\right)^2\left[\frac{1}{\epsilon}+1+3\mathbf{C}+\ln\left(-\frac{4\pi\mu^2}{k^2}\right)\right] \qquad (\,\text{DR}\,) \\ &:= \left(\frac{g}{4\pi}\right)^2\left[\sum_{s=1}^{S} C_s\,\ln\lambda_s+\frac{1}{4}+\frac{1}{2}\ln\left(-\frac{M^2}{k^2}\right)\right] \qquad (\,\text{PV}\,) \\ &:= \left(\frac{g}{4\pi}\right)^2\ln\left[-\frac{(4K)^2}{k^2}\right] \qquad (\,\text{CUT}-\text{OFF}\,) \end{aligned}$$

Remarks

1. In the present model of a left Weyl spinor minimally coupled to a gauge vector potential, no mass term can be generated by the radiative corrections in any regularisation scheme. The left-handed part of the classical kinetic term does renormalise, while its right-handed part does not undergo any radiative correction and keeps on being free. The latter has to be necessarily introduced in order to define a Feynman propagator for the massless spinor field, much like the gauge fixing term is introduced in order to invert the kinetic term of the gauge potential. The (one loop) renormalised Lagrangian for a Weyl fermion minimally coupled to a gauge vector potential has the universal—i.e., regularisation independent—form

$$\begin{aligned}
\mathcal{L}_{\,\mathrm{ren}} &= \chi_L^\dagger\,\alpha^\nu i\partial_\nu\chi_L + gA^\nu\,\chi_L^\dagger\,\alpha_\nu\chi_L - \tfrac14 F^{\mu\nu}F_{\mu\nu}\\
&+ \varphi_R^\dagger\,\alpha^\nu i\partial_\nu\varphi_R - \tfrac12(\partial\cdot A)^2 - (Z_3-1)\tfrac14 F^{\mu\nu}F_{\mu\nu}\\
&+ (Z_2-1)\chi_L^\dagger\,\alpha^\nu i\partial_\nu\chi_L + (Z_1-1)gA^\nu\,\chi_L^\dagger\,\alpha_\nu\chi_L\\
(Z_2-1)_{1-\mathrm{loop}} &= -\left(\frac{g}{4\pi}\right)^2\left[\frac{1}{\epsilon} + F_2(\epsilon, k^2/\mu^2)\right]\\
&= -\left(\frac{g}{4\pi}\right)^2\left[\sum_{s=1}^{S} C_s\,\ln\lambda_s + \widetilde{F}_2(\lambda_s, k^2/M^2)\right]\\
&= -\left(\frac{g}{4\pi}\right)^2\left\{\ln\left[-\frac{(4K)^2}{k^2}\right] + \widehat{F}_2(K^2/k^2)\right\}
\end{aligned}$$

where the customary notations have been employed. Notice that the arbitrary finite parts $F_2, \widetilde{F}_2, \widehat{F}_2$ of the countertems are analytic for $\epsilon \to 0$ and $\lambda_s, K \to \infty$, respectively, and have to be univocally fixed by the renormalisation prescription, as usual.

2. The interaction definitely preserves left chirality and scale invariance of the counterterms in the transition from the classical to the (perturbative) quantum theory: no mass coupling between the left-handed (interacting) Weyl spinor χ_L and right-handed (free) Weyl spinor φ_R can be generated by radiative loop corrections.

3. While the cut-off and dimensional regularised theory does admit a local formulation in $D=4$ or $D=2\omega$ space-time dimensions, there is no such local formulation for the Pauli–Villars regularisation. The reason is that the PV spinor propagator

$$\sum_{s=0}^{S} C_s\,S(\ell\,,\,M_s)$$

where $M_0=0\,,\;C_0=1$ while $\{\,M_s\equiv\lambda_s M\,|\,\lambda_s\gg 1\;(\,s=1,2,\dots,S\,)\,\}$, cannot be the inverse of any local differential operator of the Calderon–Zygmund type. Hence, there is no local action involving a bilinear spinor term that can produce, after a suitable inversion, the Pauli–Villars regularised spinor propagator. Although the Seeley–Schwinger–DeWitt method is not the main concern of this paper, there are no doubts that the Pauli–Villars regularisation cannot be applied to the construction of a regularised full kinetic operator for the Seeley–Schwinger–DeWitt method, nor, of course, to its inverse.

5. Majorana Massless Quantum Field

One can write a relativistic invariant field equation for a massless 2-component spinor field: to this aim, let us start from a left Weyl spinor $\psi_L\in D(\frac12,0)$ that transforms according to the $SL(2,\mathbb{C})$ matrix Λ_L. Call such a 2-component spinor field $\chi_a(x)$ $(a=1,2)$. Let us consider the Weyl spinor wave field as a classical anti-commuting field. A Majorana classical spinor field is a self-conjugated bispinor, that can be constructed, for example, out of the left-handed spinor $\chi_a(x)$ $(a = 1,2)$ as follows: namely,

$$\chi_M(x) = \left(\begin{array}{c} \chi(x) \\ -\sigma_2\chi^*(x) \end{array} \right) = \chi_M^c(x)$$

the charge conjugation rule for any classical bispinors ψ being defined by the general relationship

$$\psi^c(x) = e^{i\theta}\,\gamma^2\,\psi^*(x) \qquad\qquad (0 \leq \theta < 2\pi\,)$$

which is a discrete internal—i.e., space-time point independent—symmetry transformation. Here below we shall suitably choose $\theta = 0$. The Majorana bispinor has a right-handed lower Weyl spinor component $-\sigma_2\chi^* \in D(0,\frac{1}{2})$, albeit functional dependent, due to the charge self-conjugation constraint, in such a manner that χ_M possesses both chiralities and polarizations, at variance with its left-handed Weyl building spinor $\chi(x)$. There is another kind of self-conjugated Majorana bispinor, which can be set-up out of a right-handed Weyl building spinor $\varphi \in D(0,\frac{1}{2})$.

From the Majorana self-conjugated bispinors, one can readily construct the most general Poincaré invariant and power counting renormalisable Lagrangian. For instance, by starting from the bispinor $\chi_M(x)$ we have

$$\mathcal{L}_M = \tfrac{1}{4}\,\overline{\chi}_M(x)\,\gamma^\mu\,i\,\overleftrightarrow{\partial}_\mu\chi_M^c(x)$$

where $\alpha^\nu = \gamma_0\gamma^\nu$, while the employed notation reminds us that the upper and lower components of a Majorana bispinor can never be treated as functionally independent, even formally, due to the presence of the self-conjugation constraint. It follows that the massless Majorana action integral

$$\int \mathrm{d}^4x\,\chi_M^\dagger(x)\tfrac{1}{4}\,\alpha^\mu\gamma^2\,i\,\overleftrightarrow{\partial}_\mu\chi_M^*(x)$$

is not invariant under the overall phase transformation $\chi_M'(x) = e^{i\theta}\chi_M(x)$ of the Majorana bispinor. Hence it turns out that, as it will be further endorsed after the transition to the Majorana representation of the Dirac matrices, there is no invariant scalar charge for a Majorana spinor, which is a genuinely neutral spin $\frac{1}{2}$ field. As it will be clarified in the sequel, there is a relic continuous U(1) symmetry only for Majorana massless spinors, which drives to the existence of a conserved *pseudo-scalar charge*, the meaning of which will be better focused further on.

The massless Majorana Lagrangian in the 2-component formalism reads

$$\widehat{\mathcal{L}}_M = \tfrac{1}{2}\,\chi^\dagger(x)\,\sigma^\mu\,i\partial_\mu\chi(x) + \text{c.c.} \tag{49}$$

so that the Euler-Lagrange field equation may be written in the equivalent forms

$$i\sigma^\mu\,\partial_\mu\chi(x) = 0 \qquad\qquad i\bar{\sigma}^\mu\,\sigma_2\,\partial_\mu\chi^*(x) = 0 \tag{50}$$

which are nothing but the pair of the Weyl wave equations for both a left-handed Weyl spinor $\chi(x)$ and a right-handed Weyl spinor $\sigma_2\chi^*(x)$. This means, of course, that a massless Majorana spinor field always involves a pair of Weyl spinor fields—albeit functional dependent due to the self-conjugation constraint—with opposite chirality. As a further consequence we find that

$$\bar{\sigma}^\nu\sigma^\mu\,\partial_\nu\partial_\mu\chi(x) = \Box\chi(x) = 0$$

which means that the left-handed spinor $\chi \in D(\frac{1}{2},0)$, the building block of the self-conjugated Majorana bispinor, is actually solution of the d'Alembert wave equation. It is easy to check that the pair of Equation (50) is equivalent to the single bispinor equation

$$\alpha^\nu i\partial_\nu\chi_M(x) = 0 \tag{51}$$

where use has been made of the Dirac notation $\beta = \gamma_0$, while the Majorana Lagrangian can be recast in a further 4-component form

$$\mathcal{L}_M = \tfrac{1}{4}\,\chi_M^{\dagger}(x)\,\alpha^{\mu}\,i\,\overleftrightarrow{\partial}_{\mu}\chi_M^{c}(x) \tag{52}$$

Notice that the Majorana self-conjugated bispinor transforms under the Poincaré group as

$$\chi_M'(x') = \Lambda_{\frac{1}{2}}\,\chi_M^{c}(x) \qquad \Lambda_{\frac{1}{2}} = \begin{pmatrix} \Lambda_L & 0 \\ 0 & \Lambda_R \end{pmatrix} \tag{53}$$

with $x' = \Lambda(x + \mathrm{a})$, Λ being the Lorentz matrices in the vector representation and a^{μ} a constant space-time translation.

It turns out that, by definition, the Majorana bispinor $\chi_M(x) = \chi_M^{c}(x)$ must fulfil the self-conjugation constraint, which linearly relates the lower spinor component to the complex conjugate of the upper spinor component. Then, a representation must exist which makes the Majorana bispinor real, in such a manner that the previously introduced pair of complex variables $\chi_a \in \mathbb{C}\ (a = 1,2)$ could be replaced by the four real variables $\psi_{M,\alpha} \in \mathbb{R}\ (\alpha = 1,2,3,4)$. To obtain this real representation, we note that

$$\chi_M = \begin{pmatrix} \chi \\ -\sigma_2\,\chi^* \end{pmatrix} \qquad \chi_M^* = \begin{pmatrix} 0 & -\sigma_2 \\ \sigma_2 & 0 \end{pmatrix} \chi_M = -\gamma^2\chi_M$$

A transformation to *real* bispinor fields $\psi_M = \psi_M^*$ can be made by writing

$$\chi_M = S\,\psi_M \qquad \chi_M^* = S^*\,\psi_M = S^*\,S^{-1}\,\chi_M$$

with

$$S = \frac{\sqrt{2}}{2}\begin{pmatrix} 1 & 0 & 0 & -i \\ 0 & 1 & i & 0 \\ 0 & i & 1 & 0 \\ -i & 0 & 0 & 1 \end{pmatrix}$$

From the above relation $\psi_M = S^{-1}\chi_M = \psi_M^*$ one can immediately obtain the correspondence rule between the complex and real forms of the self-conjugated Majorana bispinor: namely,

$$\psi_{M1} = \sqrt{2}\,\Re\mathrm{e}\,\chi_1 \qquad \psi_{M2} = \sqrt{2}\,\Re\mathrm{e}\,\chi_2$$
$$\psi_{M3} = -\sqrt{2}\,\Im\mathrm{m}\,\chi_2 \quad \psi_{M4} = \sqrt{2}\,\Im\mathrm{m}\,\chi_1$$

Thus we can make use of the so called Majorana representation for the Clifford algebra which is given by the similarity transformation acting on the γ-matrices in the Weyl representation

$$\gamma_M^{\mu} \equiv S^{\dagger}\,\gamma^{\mu}\,S$$

which satisfy by direct inspection

$$\{\gamma_M^{\mu}, \gamma_M^{\nu}\} = 2\eta^{\mu\nu} \qquad \{\gamma_M^{\nu}, \gamma_M^{5}\} = 0$$
$$\gamma_M^{0} = \beta_M^{\dagger} \qquad \gamma_M^{k} = -\,\gamma_M^{k\dagger} \qquad \gamma_M^{5} = \gamma_M^{5\dagger}$$
$$\gamma_M^{\nu} = -\,\gamma_M^{\nu *} \qquad \gamma_M^{5} = -\,\gamma_M^{5*}$$

The result is that, at the place of a complex self-conjugated bispinor, which has been constructed out of a left-handed Weyl spinor, one can safely and more suitably employ a real Majorana bispinor: namely,

$$\chi_M(x) = \chi_M^{c}(x) \qquad \leftrightarrow \qquad \psi_M(x) = S^{\dagger}\chi_M(x) = \psi_M^*(x)$$

A quite analogous construction can obviously be made, had we started from a right-handed Weyl spinor $\varphi \in D(0, \frac{1}{2})$. Then, the massless Majorana Lagrangian and the ensuing wave field equation take the manifestly real forms

$$\mathcal{L}_M = \tfrac{1}{4}\, \psi_M^\top(x)\, \alpha_M^\nu\, i \overleftrightarrow{\partial}_\nu\, \psi_M(x)$$
$$i\partial\!\!\!/_M \psi_M(x) = 0 \qquad\qquad \psi_M(x) = \psi_M^*(x)$$
$$\alpha_M^\nu = \gamma_M^0\, \gamma_M^\nu \qquad \alpha_M^0 = \mathbb{I} \qquad \beta_M \equiv \gamma_M^0$$

It turns out that, from the manifestly real form of the Majorana Lagrangian, the only relic internal symmetries of the massless Majorana's action integral are the discrete $\mathbb{Z}_2$ symmetry, i.e., $\psi_M(x) \longmapsto -\,\psi_M(x)$, and a further continuous symmetry under the chiral U(1) group

$$\psi_M(x) \quad \mapsto \quad \psi_M'(x) = \exp\left\{\pm\, i\theta\, \gamma_M^5\right\} \psi_M(x) \qquad\qquad (\,0 \le \theta < 2\pi\,) \tag{54}$$

the imaginary unit being convenient to keep the reality of the transformed Majorana bispinor. From Nöther theorem, we get the corresponding real current, which satisfies the continuity equation

$$j_5^\mu(x) = \tfrac{1}{2}\, \psi_M^\top(x)\, \alpha_M^\mu\, i\gamma_M^5\, \psi_M(x) \qquad\qquad \partial \cdot j_5(x) = 0 \tag{55}$$

as well as the ensuing conserved pseudo-scalar charge

$$\pm\, Q_5 = \pm\, \frac{1}{2} \int \mathrm{d}\mathbf{x}\, \psi_M^\top(t, \mathbf{x})\, i\gamma_M^5 \psi_M(t, \mathbf{x}) \qquad\qquad \dot{Q}_5 = 0$$

the overall $\pm$ sign being conventional and irrelevant. For anti-commuting Grassmann-valued functions we get

$$\pm\, Q_5 = \pm \int \mathrm{d}\mathbf{x}\ [\, \psi_{M,1}(t, \mathbf{x}) \psi_{M,4}(t, \mathbf{x}) - \psi_{M,2}(t, \mathbf{x}) \psi_{M,3}(t, \mathbf{x})\,]$$

the integrated quantity being nothing but than $-\,i\chi^\dagger(x)\chi(x)$ as expected.

The chiral symmetry (54) has been already discussed at the end of Section 2. In Section 8, we will see that the conservation law (55) at the quantum level is violated by an anomaly.

It is interesting to set up the spin-states of the massless Majorana real spinor field, because they sensibly differ from the usual and well-known Dirac spin states (the spin states for Weyl spinor are summarized in Appendix A). Let us start from the constant eigenvectors $\xi_\pm$ of the chiral matrix γ_M^5 in the Majorana representation which do indeed satisfy by definition

$$\gamma_M^5\, \xi_\pm = \pm\, \xi_\pm$$

Then, for example, we can suitably define the Majorana massless spin-states to be

$$w_r(\mathbf{p}) \equiv p\!\!\!/_M\, \xi_r / 2\wp \qquad\quad (\, r = \pm,\ p_0 = \wp \equiv |\mathbf{p}|\,)$$

and the corresponding plane wave functions

$$w_{\mathbf{p},r}(x) \equiv [\, (2\pi)^3 2\wp\,]^{-\frac{1}{2}}\, w_r(\mathbf{p})\, \exp\{-\, i\wp t + i\mathbf{p} \cdot \mathbf{x}\}$$

The above introduced Majorana massless spin-states are complex and the related plane waves turn out to be a pair of degenerate eigenstates positive frequency or energy solutions of the massless Majorana wave equation $i\partial_\mu \gamma_M^\mu w_{\mathbf{p},r}(x) = 0\ (\, r = +, -\,)$. It appears that the pair of orthogonal negative energy spin-states and plane waves functions are nothing but the complex conjugates of the former ones.

The transition to the quantum theory is performed as usual by means of the creation annihilation operators $a_{\mathbf{p},r}$ and $a^{\dagger}_{\mathbf{p},s}$ which satisfy the canonical anti-commutation relations

$$\{a_{\mathbf{p},r}\,,\,a_{\mathbf{q},s}\} = 0 = \{a^{\dagger}_{\mathbf{p},r}\,,\,a^{\dagger}_{\mathbf{q},s}\} \qquad \{a_{\mathbf{p},r}\,,\,a^{\dagger}_{\mathbf{q},s}\} \;=\; \delta_{rs}\,\delta\,(\mathbf{p}-\mathbf{q})$$

with $\mathbf{p},\,\mathbf{q}\in\mathbb{R}^3$, $r,s=\pm$, so that the operator valued tempered distribution for the Majorana quantum spinor field takes the form

$$\psi_M(x) = \sum_{\mathbf{p},r}\left[\,a_{\mathbf{p},r}\,w_{\mathbf{p},r}(x) + a^{\dagger}_{\mathbf{p},r}\,w^{*}_{\mathbf{p},r}(x)\,\right] = \psi^{c}_{M}(x)$$

where use has been made of the shorthand notation $\sum_{\mathbf{p},r}\equiv\int \mathrm{d}\mathbf{p}\sum_{r=\pm}$. Moreover, we readily get the useful expansion for the adjoint quantum fields

$$\overline{\psi}_M(x) = \sum_{\mathbf{q},s}\left[\,a^{\dagger}_{\mathbf{q},s}\,\overline{w}_{\mathbf{q},s}(x) + a_{\mathbf{q},s}\,\overline{w}^{*}_{\mathbf{q},s}(x)\,\right]$$

in which

$$\overline{w}_{\mathbf{p},r}(x) \equiv [\,(2\pi)^3 2\wp\,]^{-\frac{1}{2}}\,[\,w_r^{\top}(\mathbf{p})\,]^{*}\,\beta_M\,\exp\{\,i\wp t - i\mathbf{p}\cdot\mathbf{x}\}$$

From the orthogonality relations for spin-states and plane waves, we get the expressions for the energy-momentum and helicity observables in terms of normal ordered products of operators

$$P_\mu = \frac{i}{4}\int \mathrm{d}\mathbf{x}\,:\overline{\psi}_M(x)\,\beta_M\,\overleftrightarrow{\partial}_\mu\,\psi_M(x): = \sum_{\mathbf{p},r}\,p_\mu\,a^{\dagger}_{\mathbf{p},r}\,a_{\mathbf{p},r} \qquad (\,p_0=\wp\,)$$

$$\mathrm{h} = \int_{-\infty}^{\infty}\mathrm{d}y\,:\tfrac{1}{2}\,\psi_M(t,y)\,\Sigma_{M,2}\,\psi_M(t,y):=\tfrac{1}{2}\int_{-\infty}^{\infty}\mathrm{d}p\,\left[\,a^{\dagger}_{p,+}\,a_{p,+} - a^{\dagger}_{p,-}\,a_{p,-}\,\right]$$

where a 1D motion along the Oy-axis has been referred to, e.g., to select the helicity operator.

$$Q_5 = \tfrac{1}{2}\int \mathrm{d}\mathbf{x}\,:\overline{\psi}_M(t,\mathbf{x})\beta_M\gamma^5_M\psi(t,\mathbf{x}): = \int \mathrm{d}\mathbf{p}\,\left(a^{\dagger}_{\mathbf{p},+}\,a_{\mathbf{p},+} - a^{\dagger}_{\mathbf{p},-}\,a_{\mathbf{p},-}\right) \equiv \nu_M$$

whence it follows that the above introduced pseudo-scalar charge Q_5 is nothing but the quantum counterpart of the Atiah–Singer index for a Majorana spinor field.

Hence, it appears that the 1-particle states $a^{\dagger}_{\mathbf{p},\pm}\,|0\rangle$ represent neutral Majorana particles with energy-momentum $p^{\mu}=(\wp,\,\mathbf{p})$ and positive/negative helicity.

It is worthwhile to remark and gather that, as explicitly shown above, the quanta of the massless spin $\frac{1}{2}$ Majorana field are on the light-cone, neutral—i.e., particle and anti-particle actually coincide—and with two polarization states, so that they have really nothing to share with the quanta of a Weyl field, but living on the light-cone, the latter being charged, the particle and antiparticle carrying one single and opposite polarization. In a sense, the mechanical properties of the spin $\frac{1}{2}$ massless Majorana quanta are close and similar to those ones of a spin 1 photons, apart from the interactions.

Finally, from the massless Majorana Lagrangian and related wave equation, one can immediately realize that the causal Green function for the Majorana massless neutral spinor field is the very same as for a massless Dirac charged quantum field: namely,

$$\langle\,0\,|\,T\,\psi_M(x)\,\overline{\psi}_M(y)\,|\,0\,\rangle = S_M(x-y) = \frac{i}{(2\pi)^4}\int \mathrm{d}^4p\,\frac{\not{p}_M}{p^2+i\varepsilon}\,\exp\{-\,ip\cdot x\}$$

the massless limit being smooth, which satisfies $i\not{\partial}_M S_M(x-y) = i\delta(x-y)$.

The second part of this review is devoted to gauge anomalies. Spin states are indispensable for the S matrix elements, but do not play a direct role in the calculation of anomalies. For them we need the fermion determinant and its variations. A formal manipulation of the path integral shows that the determinant of the Dirac operator for a Dirac spinor is the square of the fermion determinant for a Majorana spinor. One can take the square root of the Dirac determinant as the definition of the fermion determinant (the functional integral) of a Majorana fermion. This formal procedure turns out to be correct, as one can check by comparing with the perturbative approach. In Section 8, we present one of such checks: based on the perturbative results for Weyl fermions and using the representation of a Majorana fermion in terms a Weyl fermion and its Lorentz covariant conjugate (see comment after Equation (12)), we show there that the anomalies for a Majorana fermion are the same as those of a Dirac fermion, with half coefficient.)

6. Consistent Gauge Anomalies for Weyl Fermions

As explained in the introduction, anomalies are one of the main topics where Dirac and Weyl fermions split significantly. In the present and forthcoming sections we aim to recalculate all the anomalies (chiral and trace) of Weyl, Dirac and Majorana fermions coupled to gauge potentials, with the basic method of Feynman diagrams. Most results are supposedly well-known. Our purpose is to collect them all in order to highlight their reciprocal relations.

Let us consider the classical action integral for a right-handed Weyl fermion coupled to an external gauge field $V_\mu = V_\mu^a T^a$, T^a being Hermitean generators, $[T^a, T^b] = i f^{abc} T^c$ (in the Abelian case $T = 1, f = 0$) in a fundamental representation of e.g., SU(N): namely,

$$S_R[V] = \int d^4x\, i\, \overline{\psi_R}\, (\partial\!\!\!/ - i V\!\!\!/)\, \psi_R. \tag{56}$$

This action is invariant under the gauge transformation $\delta V_\mu = D_\mu \lambda \equiv \partial_\mu \lambda - i[V_\mu, \lambda]$, which implies the conservation of the non-Abelian current $J^a_{R\mu} = \bar\psi_R \gamma_\mu T^a \psi_R$, i.e.

$$\nabla \cdot J^a_R \equiv (\partial^\mu \delta^{ac} + f^{abc} V^{b\mu}) J^c_{R\mu} = 0. \tag{57}$$

The quantum effective action for this theory is given by the generating functional of the connected Green functions of such currents in the presence of the source $V^{a\mu}$

$$\begin{aligned} W[V] &= W[0] \\ &+ \sum_{n=1}^{\infty} \frac{i^{n-1}}{n!} \int \prod_{i=1}^{n} d^4x_i V^{a_i \mu_i}(x_i)\, \langle 0|\mathcal{T} J^{a_1}_{R\mu_1}(x_1) \dots J^{a_n}_{R\mu_n}(x_n)|0\rangle_c \end{aligned} \tag{58}$$

and the full 1-loop 1-point function of $J^a_{R\mu}$ is

$$\langle\langle J^a_{R\mu}(x) \rangle\rangle = \frac{\delta W[V]}{\delta V^{a\mu}(x)} = \sum_{n=0}^{\infty} \frac{i^n}{n!} \int \prod_{i=1}^{n} d^4x_i V^{a_i\mu_i}(x_i) \langle 0|\mathcal{T} J^a_{R\mu}(x) J^{a_1}_{R\mu_1}(x_1) ... J^{a_n}_{R\mu_n}(x_n)|0\rangle_c \tag{59}$$

Our purpose here is to calculate the odd parity anomaly of the divergence $\nabla \cdot \langle\langle J^a_R \rangle\rangle$. As is well-known, the first nontrivial contribution to the anomaly comes from the divergence of the three-point function in the RHS of (59). For simplicity, we will denote it $\langle \partial \cdot J_R\, J_R\, J_R \rangle$. Below we will evaluate it in some detail as a sample for the remaining calculations.

6.1. The Calculation

Let us start with dimensional regularisation. The fermion propagator is $\frac{i}{p\!\!\!/}$ and the vertex $i\gamma_\mu P_R T^a$. The Fourier transform of the three currents amplitude $\langle J_R\, J_R\, J_R \rangle$ is given by

$$\begin{aligned} \widetilde{F}^{(R)abc}_{\mu\lambda\rho}(k_1, k_2) &= \int \frac{d^4p}{(2\pi)^4} \mathrm{Tr}\left\{ \frac{1}{p\!\!\!/} \frac{1-\gamma_5}{2} \gamma_\lambda T^b \frac{1}{p\!\!\!/ - k\!\!\!/_1} \frac{1-\gamma_5}{2} \gamma_\rho T^c \frac{1}{p\!\!\!/ - q\!\!\!/} \frac{1-\gamma_5}{2} \gamma_\mu T^a \right\} \\ &\equiv \mathrm{Tr}(T^a T^b T^c) \widetilde{F}^{(R)}_{\mu\lambda\rho}(k_1, k_2) \end{aligned} \tag{60}$$

where $q = k_1 + k_2$. The relevant Feynman diagram is shown in Figure 1.

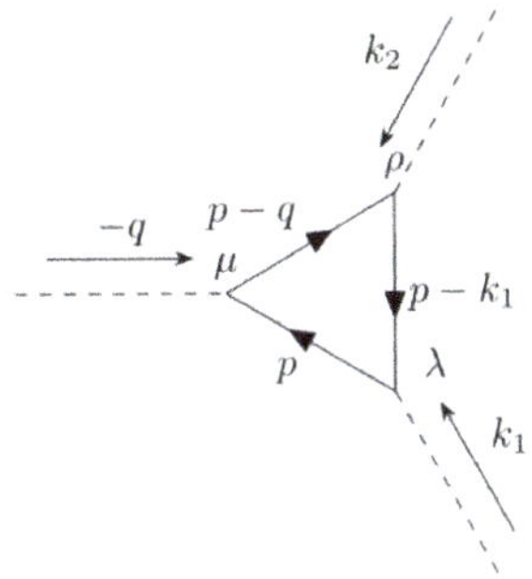

Figure 1. The Feynman diagram corresponding to $\widetilde{F}^{(R)abc}_{\mu\lambda\rho}(k_1, k_2)$.

From now on we focus on the Abelian part $\widetilde{F}^{(R)}_{\mu\lambda\rho}(k_1, k_2)$. We dimensionally regularise it by introducing δ additional dimensions and corresponding momenta ℓ_μ, $\mu = 0, \ldots, 3 + \delta$, with the properties

$$\not{\ell}\not{p} + \not{p}\not{\ell} = 0, \qquad [\not{\ell}, \gamma_5] = 0, \qquad \not{p}^2 = p^2, \qquad \not{\ell}^2 = -\ell^2$$
$$\mathrm{tr}(\gamma_\mu\gamma_\nu\gamma_\lambda\gamma_\rho\gamma_5) = -2^{2+\frac{\delta}{2}}\, i\,\epsilon_{\mu\nu\lambda\rho},$$

so the relevant expression to be calculated is

$$q^\mu \widetilde{F}^{(R)}_{\mu\lambda\rho}(k_1, k_2) = \int \frac{d^4p d^\delta \ell}{(2\pi)^{4+\delta}} \mathrm{tr}\left\{\frac{1}{\not{p} + \not{\ell}}\frac{1-\gamma_5}{2}\gamma_\lambda \frac{1}{\not{p} + \not{\ell} - \not{k}_1}\frac{1-\gamma_5}{2}\gamma_\rho \frac{1}{\not{p} + \not{\ell} - \not{q}}\frac{1-\gamma_5}{2}\not{q}\right\}$$
$$= \int \frac{d^4p d^\delta \ell}{(2\pi)^{4+\delta}} \mathrm{tr}\left\{\frac{\not{p}}{p^2 - \ell^2}\gamma_\lambda \frac{\not{p} - \not{k}_1}{(p-k_1)^2 - \ell^2}\gamma_\rho \frac{\not{p} - \not{q}}{(p-q)^2 - \ell^2}\frac{1-\gamma_5}{2}\not{q}\right\}$$
$$\equiv \widetilde{F}^{(R)}_{\lambda\rho}(k_1, k_2, \delta). \tag{61}$$

Now we focus on the odd part and work out the gamma traces:

$$\widetilde{F}^{(R,odd)}_{\lambda\rho}(k_1, k_2, \delta) = -2^{1+\frac{\delta}{2}}\, i\epsilon_{\mu\nu\lambda\rho} \int \frac{d^4p d^\delta \ell}{(2\pi)^{4+\delta}} \frac{\left(p^2 q^\mu + (q^2 - 2p\cdot q)p^\mu\right)\left(p^\nu - k_1^\nu\right)}{(p^2 - \ell^2)((p-k_1)^2 - \ell^2)((p-q)^2 - \ell^2)}. \tag{62}$$

Let us write the numerator on the RHS as follows:

$$p^2 q^\mu + (q^2 - 2p\cdot q)p^\mu = -(p^2 - \ell^2)(p-q)^\mu + ((p-q)^2 - \ell^2)p^\mu + \ell^2 q^\mu. \tag{63}$$

Then (62) can be rewritten as

$$\widetilde{F}^{(R,odd)}_{\lambda\rho}(k_1, k_2, \delta) = -2^{1+\frac{\delta}{2}}\, i\epsilon_{\mu\nu\lambda\rho} \int \frac{d^4p d^\delta \ell}{(2\pi)^{4+\delta}} \left\{\ell^2 \frac{q^\mu\left(p^\nu - k_1^\nu\right)}{(p^2 - \ell^2)((p-k_1)^2 - \ell^2)((p-q)^2 - \ell^2)} + \frac{(q-p)^\mu (p-k_1)^\nu}{((p-k_1)^2 - \ell^2)((p-q)^2 - \ell^2)} + \frac{p^\mu (p-k_1)^\nu}{(p^2 - \ell^2)((p-k_1)^2 - \ell^2)}\right\} \tag{64}$$

$$= -2^{1+\frac{\delta}{2}}\, i\epsilon_{\mu\nu\lambda\rho} \int \frac{d^4p d^\delta \ell}{(2\pi)^{4+\delta}} \left\{\ell^2 \frac{q^\mu\left(p^\nu - k_1^\nu\right)}{(p^2 - \ell^2)((p-k_1)^2 - \ell^2)((p-q)^2 - \ell^2)} - \frac{(p-k_2)^\mu p^\nu}{(p^2 - \ell^2)((p-k_2)^2 - \ell^2)} + \frac{p^\mu (p-k_1)^\nu}{(p^2 - \ell^2)((p-k_1)^2 - \ell^2)}\right\}. \tag{65}$$

The last two terms do not contribute because of the antisymmetric ϵ tensor, as one can easily see by introducing a Feynman parameter. The first term can be easily evaluated by introducing two Feynman parameters x and y, and making the shift $p \to p + (x+y)k_1 + yk_2$,

$$\widetilde{F}^{(R,odd)}_{\lambda\rho}(k_1,k_2,\delta) = -2^{2+\frac{\delta}{2}}\, i\epsilon_{\mu\nu\lambda\rho} \int_0^1 dx \int_0^{1-x} dy \int \frac{d^4p d^\delta \ell}{(2\pi)^{4+\delta}} \ell^2 \frac{q^\mu (p^\nu + (x+y-1)k_1 + yk_2)^\nu)}{(p^2 - \ell^2 + \Delta(x,y))^3} \tag{66}$$

where $\Delta = (x+y)(1-x-y)k_1^2 + y(1-y)k_2^2 + 2y(1-x-y)k_1 \cdot k_2$. Now we make a Wick rotation on the integration momentum, $p^0 \to ip^0$, and the same on k_1, k_2 (although we stick to the same symbols).

Then, using

$$\int \frac{d^4p}{(2\pi)^4} \int \frac{d^\delta \ell}{(2\pi)^\delta} \frac{\ell^2}{(p^2 + \ell^2 + \Delta)^3} = -\frac{1}{2(4\pi)^2} \tag{67}$$

and taking the limit $\delta \to 0$, we find

$$\widetilde{F}^{(R,odd)}_{\lambda\rho}(k_1,k_2) = \frac{2}{(4\pi)^2} \epsilon_{\mu\nu\lambda\rho} k_1^\mu k_2^\nu \int_0^1 dx \int_0^{1-x} dy\, (1-x) = \frac{1}{24\pi^2} \epsilon_{\mu\nu\lambda\rho} k_1^\mu k_2^\nu. \tag{68}$$

We must add the cross term (for $\lambda \leftrightarrow \rho$ and $k_1 \leftrightarrow k_2$), so that the total result is

$$\widetilde{F}^{(R,odd)}_{\lambda\rho}(k_1,k_2) + \widetilde{F}^{(R,odd)}_{\rho\lambda}(k_2,k_1) = \frac{1}{12\pi^2} \epsilon_{\mu\nu\lambda\rho} k_1^\mu k_2^\nu. \tag{69}$$

In order to return to configuration space we have to insert this result into (59). We consider here, for simplicity, the Abelian case. We have

$$\begin{aligned} \partial^\mu \langle\!\langle J_{R\mu}(x) \rangle\!\rangle &= \int \frac{d^4q}{(2\pi)^4} e^{-iqx} (-iq^\mu) \langle\!\langle \tilde{J}_{R\mu}(q) \rangle\!\rangle = \frac{i}{2} \int \frac{d^4q}{(2\pi)^4} \int \frac{d^4k_1}{(2\pi)^4} \int \frac{d^4k_2}{(2\pi)^4} \int d^4y d^4z \\ &\times q^\mu\, e^{i(k_1y + k_2z - qx)} \left(\widetilde{F}^{(R,odd)}_{\lambda\rho}(k_1,k_2) + \widetilde{F}^{(R,odd)}_{\rho\lambda}(k_2,k_1) \right) V^\lambda(y) V^\rho(z). \end{aligned} \tag{70}$$

After a Wick rotation, we can replace (69) inside the integrals

$$\begin{aligned} \partial^\mu \langle\!\langle J_{R\mu}(x) \rangle\!\rangle &= -\frac{1}{24\pi^2} \int \frac{d^4q\, d^4k_1\, d^4k_2}{(2\pi)^{12}} \int d^4y d^4z\, e^{i(qx - k_1y - k_2z)} \delta(q - k_1 - k_2) \epsilon_{\mu\nu\lambda\rho} k_1^\mu k_2^\nu V^\lambda(y) V^\rho(z) \\ &= -\frac{1}{24\pi^2} \int \frac{d^4k_1}{(2\pi)^4} \int \frac{d^4k_2}{(2\pi)^4} \int d^4y d^4z\, e^{ik_1(x-y)} e^{-ik_2(x-z)} \epsilon_{\mu\nu\lambda\rho}\, \partial^\mu V^\lambda(y) \partial^\nu V^\rho(z) \\ &= \frac{1}{24\pi^2} \epsilon_{\mu\nu\lambda\rho}\, \partial^\mu V^\nu(x) \partial^\lambda V^\rho(x). \end{aligned} \tag{71}$$

The same result can be obtained with the Pauli–Villars regularisation, see Appendix B.

6.2. Comments

Equation (71) is the consistent gauge anomaly of a right-handed Weyl fermion coupled to an Abelian vector field $V_\mu(x)$. It is well known that the consistent anomaly (71) destroys the consistency of the Abelian gauge theory. As a matter of fact the Lorentz invariant quantum theory of a gauge vector field unavoidably involves a Fock space of states with indefinite norm. Now, in order to select a physical Hilbert subspace of the Fock space, a subsidiary condition is necessary. In the Abelian case, when the fermion current satisfies the continuity equation the equations of motion lead to $\Box(\partial \cdot V) = 0$, so that a subspace of states of non-negative norm can be selected through the auxiliary condition

$$\partial \cdot V^{(-)}(x) \,|\, \mathrm{phys} \,\rangle = 0$$

$V^{(-)}(x)$ being the annihilation operator, the positive frequency part of a d'Alembert quantum field. On the contrary , in the present chiral model we find

$$\Box(\partial \cdot V) = -\frac{1}{3}\left(\frac{1}{4\pi}\right)^2 F_*^{\mu\nu} F_{\mu\nu} \neq 0$$

in such a manner that nobody knows how to select a physical subspace of states with non-negative norm, if any, where a unitary restriction of the collision operator S could be defined.

Another way of seeing the problem created by the consistent anomaly is to remark that, for instance, $J_{R\mu}$ couples minimally to V^μ at the fermion-fermion-gluon vertex. Unitarity and renormalisability rely on the Ward identity that guarantees current conservation at any such vertex. This is impossible in the presence of a consistent anomaly.

The consistent anomaly in the non-Abelian case would require the calculation of at least the four current correlators, but it can be obtained in a simpler way from the Abelian case using the Wess–Zumino consistency conditions. In the non-Abelian case the three-point correlators are multiplied by

$$\mathrm{Tr}(T^a T^b T^c) = \frac{1}{2}\mathrm{Tr}(T^a[T^b, T^c]) + \frac{1}{2}\mathrm{Tr}(T^a\{T^b, T^c\}) = f^{abc} + d^{abc} \tag{72}$$

where the normalisation used is $\mathrm{Tr}(T^a T^b) = 2\delta^{ab}$. Since the three-point function is the sum of two equal pieces with $\lambda \leftrightarrow \rho, k_1 \leftrightarrow k_2$, the first term in the RHS of (72) drops out and only the second remains. For the right-handed current $J^a_{R\mu}$ we have

$$\nabla \cdot \langle\langle J^a_R \rangle\rangle = \frac{1}{24\pi^2}\varepsilon_{\mu\nu\lambda\rho}\mathrm{Tr}\left[T^a \partial^\mu \left(V^\nu \partial^\lambda V^\rho + \frac{i}{2} V^\nu V^\lambda V^\rho\right)\right]. \tag{73}$$

The previous results are well-known (for a short introduction to consistency and covariance applied to anomalies, see Appendix C). However, they do not tell the whole story about gauge anomalies in a theory of Weyl fermions. To delve into this we have to enlarge the parameter space by coupling the fermions to an additional potential, namely to an axial vector field.

7. The *V*-*A* Anomalies

The action of a Dirac fermion coupled to a vector V_μ and an axial potential A_μ (for simplicity we consider only the Abelian case) is

$$S[V, A] = \int d^4x\, i\,\overline{\psi}\,(\partial\!\!\!/ - iV\!\!\!\!/ - iA\!\!\!/\gamma_5)\,\psi. \tag{74}$$

The generating functional of the connected Green functions is

$$\begin{aligned} W[V, A] &= W[0,0] + \sum_{n,m=1}^{\infty} \frac{i^{n+m-1}}{n!m!} \int \prod_{i=1}^{n} d^4x_i\, V^{\mu_i}(x_i) \prod_{j=1}^{m} d^4y_j A^{\nu_j}(y_j) \\ &\quad \times \langle 0|T J_{\mu_1}(x_1) \ldots J_{\mu_n}(x_n) J_{5\nu_1}(y_1) \ldots J_{5\nu_m}(x_m)|0\rangle_c. \end{aligned} \tag{75}$$

We can extract the full one-loop one-point function for two currents: the vector current $J_\mu = \bar{\psi}\gamma_\mu\psi$

$$\begin{aligned} \langle\langle J_\mu(x) \rangle\rangle &= \frac{\delta W[V, A]}{\delta V^\mu(x)} = \sum_{n,m=0}^{\infty} \frac{i^{n+m}}{n!m!} \int \prod_{i=1}^{n} d^4x_i\, V^{\mu_i}(x_i) \prod_{j=1}^{m} d^4y_j A^{\nu_j}(y_j) \\ &\quad \times \langle 0|T J_\mu(x) J_{\mu_1}(x_1) \ldots J_{\mu_n}(x_n) J_{5\nu_1}(y_1) \ldots J_{5\nu_m}(x_m)|0\rangle_c \end{aligned} \tag{76}$$

and the axial current $J_\mu = \bar{\psi}\gamma_\mu\gamma_5\psi$

$$\begin{aligned}\langle\langle J_{5\mu}(x)\rangle\rangle &= \frac{\delta W[V,A]}{\delta A^\mu(x)} = \sum_{n,m=0}^{\infty} \frac{i^{n+m}}{n!m!} \int \prod_{i=1}^{n} d^4x_i\, V_{\mu_i}(x_i) \prod_{j=1}^{m} d^4y_j A^{\nu_j}(y_j) \\ &\times \langle 0|\mathcal{T} J_{5\mu}(x) J_{\mu_1}(x_1)\dots J_{\mu_n}(x_n) J_{5\nu_1}(y_1)\dots J_{5\nu_m}(x_m)|0\rangle_c. \end{aligned} \tag{77}$$

These currents are conserved except for possible anomaly contributions. The aim of this section is to study the continuity equations for these currents, that is to compute the 4-divergences of the correlators on the RHS of (76) and (77). For the same reason explained above we focus on the three current correlators: they are all we need in the Abelian case (and the starting point to compute the full anomaly expression by means of the Wess–Zumino consistency conditions in the non-Abelian case). For $\partial \cdot J(x)$, the first relevant contributions are

$$\begin{aligned}\partial^\mu \langle\langle J_\mu(x)\rangle\rangle &= -\Big(\frac{1}{2}\int d^4x_1 d^4x_2 V^{\mu_1}(x_1)V^{\mu_2}(x_2)\partial^\mu \langle 0|\mathcal{T} J_\mu(x)J_{\mu_1}(x_1)J_{\mu_2}(x_2)|0\rangle \\ &+ \int d^4x_1 d^4y_1 V^{\mu_1}(x_1)A^{\nu_1}(y_1)\partial^\mu\langle 0|\mathcal{T} J_\mu(x)J_{\mu_1}(x_1)J_{5\nu_1}(y_1)|0\rangle \\ &+\frac{1}{2}\int d^4y_1 d^4y_2 A^{\nu_1}(y_1)A^{\nu_2}(y_2)\partial^\mu\langle 0|\mathcal{T} J_\mu(x)J_{5\nu_1}(y_1)J_{5\nu_2}(y_2)|0\rangle\Big) \end{aligned} \tag{78}$$

and for $\partial^\mu J_{5\mu}(x)$

$$\begin{aligned}\partial^\mu \langle\langle J_{5\mu}(x)\rangle\rangle &= -\Big(\frac{1}{2}\int d^4x_1 d^4x_2 V^{\mu_1}(x_1)V^{\mu_2}(x_2)\partial^\mu \langle 0|\mathcal{T} J_{5\mu}(x)J_{\mu_1}(x_1)J_{\mu_2}(x_2)|0\rangle \\ &+ \int d^4x_1 d^4y_1 V^{\mu_1}(x_1)A^{\nu_1}(y_1)\partial^\mu\langle 0|\mathcal{T} J_{5\mu}(x)J_{\mu_1}(x_1)J_{5\nu_1}(y_1)|0\rangle \\ &+\frac{1}{2}\int d^4y_1 d^4y_2 A^{\nu_1}(y_1)A^{\nu_2}(y_2)\partial^\mu\langle 0|\mathcal{T} J_{5\mu}(x)J_{5\nu_1}(y_1)J_{5\nu_2}(y_2)|0\rangle\Big). \end{aligned} \tag{79}$$

Since we are interested in odd parity anomalies, the only possible contribution to (78) is from the term in the second line, which we denote concisely $\langle \partial \cdot J\, J\, J_5\rangle$. As for (79), there are two possible contributions from the first and third lines, i.e., $\langle \partial \cdot J_5\, J\, J\rangle$ and $\langle \partial \cdot J_5\, J_5\, J_5\rangle$. Below we report the results for the corresponding amplitudes, obtained with dimensional regularisation.

The amplitude for $\langle \partial \cdot J_5\, J\, J\rangle$ is

$$q^\mu \widetilde{F}^{(5)}_{\mu\lambda\rho}(k_1,k_2) = \int \frac{d^4p\, d^\delta \ell}{(2\pi)^{4+\delta}} \mathrm{tr}\left\{\frac{1}{\not p + \not \ell}\gamma_\lambda \frac{1}{\not p + \not\ell - \not k_1}\gamma_\rho \frac{1}{\not p + \not \ell - \not q} \not q \gamma_5\right\}. \tag{80}$$

The relevant Feynman diagram is shown in Figure 2.

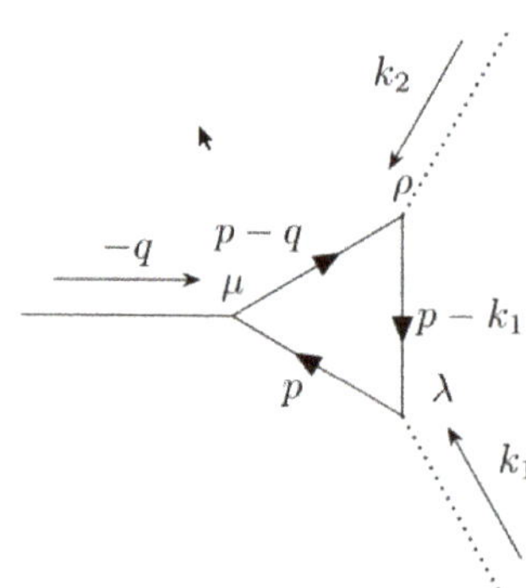

Figure 2. The Feynman diagram corresponding to $\widetilde{F}^{(5)}_{\mu\lambda\rho}(k_1,k_2)$.

Adding the cross contribution one gets

$$q^{\mu}\left(\widetilde{F}^{(5)}_{\mu\lambda\rho}(k_1,k_2)+T^{(5)}_{\mu\rho\lambda}(k_2,k_1)\right)=\frac{1}{2\pi^2}\epsilon_{\mu\nu\lambda\rho}k_1^{\mu}k_2^{\nu}. \tag{81}$$

The amplitude for $\langle\partial\cdot J_5\,J_5\,J_5\rangle$ is given by

$$\begin{aligned} q^{\mu}\widetilde{F}^{(555)}_{\mu\lambda\rho}(k_1,k_2) &= \int\frac{d^4p d^{\delta}\ell}{(2\pi)^{4+\delta}}\mathrm{tr}\left\{\frac{1}{\not p+\not\ell}\gamma_\lambda\gamma_5\frac{1}{\not p+\not\ell-\not k_1}\gamma_\rho\gamma_5\frac{1}{\not p+\not\ell-\not q}\not q\gamma_5\right\}\\ &= -2^{2+\frac{\delta}{2}}i\epsilon_{\mu\nu\lambda\rho}\int\frac{d^4p d^{\delta}\ell}{(2\pi)^{4+\delta}}\,\ell^2\,\frac{q^{\mu}(3p^{\nu}-k_1^{\nu})}{(p^2-\ell^2)((p-k_1)^2-\ell^2)((p-q)^2-\ell^2)}\\ &\quad-\int\frac{d^4p d^{\delta}\ell}{(2\pi)^{4+\delta}}\frac{\mathrm{tr}(\not q\not p\gamma_\lambda(\not p-\not k_1)\gamma_\rho(\not p-\not q)\gamma_5)}{(p^2-\ell^2)((p-k_1)^2-\ell^2)((p-q)^2-\ell^2)} \end{aligned} \tag{82}$$

The first line in the last expression, after introducing the Feynman parameters x and y and shifting p as usual, yields a factor $\int_0^1 dx\int_0^{1-x}dy\,(1-3x)=0$, so it vanishes. The last line is $2\times\widetilde{F}^{(R,odd)}_{\lambda\rho}(k_1,k_2,\delta)$, cf. (61) and (62). Therefore, using (69), we get

$$q^{\mu}\left(\widetilde{F}^{(555)}_{\mu\lambda\rho}(k_1,k_2)+\widetilde{F}^{(555)}_{\mu\rho\lambda}(k_2,k_1)\right)=\frac{1}{6\pi^2}\epsilon_{\mu\nu\lambda\rho}k_1^{\mu}k_2^{\nu}. \tag{83}$$

Finally the amplitude for $\langle\partial\cdot J\,J\,J_5\rangle$ is

$$q^{\mu}\widetilde{F}^{(5')}_{\mu\lambda\rho}(k_1,k_2) = \int\frac{d^4p d^{\delta}\ell}{(2\pi)^{4+\delta}}\mathrm{tr}\left\{\frac{1}{\not p+\not\ell}\gamma_\lambda\frac{1}{\not p+\not\ell-\not k_1}\gamma_\rho\gamma_5\frac{1}{\not p+\not\ell-\not q}\not q\right\}=0. \tag{84}$$

All the above results have been obtained also with PV regularisation.

Plugging in these results in (76) and (77) we find

$$\partial^{\mu}\langle\langle J_{\mu}(x)\rangle\rangle=0 \tag{85}$$

and

$$\partial^{\mu}\langle\langle J_{5\mu}(x)\rangle\rangle=\frac{1}{4\pi^2}\epsilon_{\mu\nu\lambda\rho}\left(\partial^{\mu}V^{\nu}(x)\partial^{\lambda}V^{\rho}(x)+\frac{1}{3}\partial^{\mu}A^{\nu}(x)\partial^{\lambda}A^{\rho}(x)\right) \tag{86}$$

which is Bardeen's result [19], in the Abelian case. From (86) we can derive the covariant chiral anomaly by setting $A_{\mu}=0$, then

$$\partial^{\mu}\langle\langle J_{5\mu}(x)\rangle\rangle=\frac{1}{4\pi^2}\epsilon_{\mu\nu\lambda\rho}\,\partial^{\mu}V^{\nu}(x)\partial^{\lambda}V^{\rho}(x). \tag{87}$$

Of course this is nothing but (81). For the $J_{5\mu}(x)$ current is obtained by differentiating the action with respect to $A_{\mu}(x)$ and its divergence leads to the covariant anomaly.

Some Conclusions

Let us recall that in the collapsing limit $V\to V/2, A\to V/2$ in the action (74) we recover the theory of a right-handed Weyl fermion (with the addition of a free left-handed part, as explained at length above). Now $J_{\mu}(x)=J_{R\mu}(x)+J_{L\mu}(x)$ and $J_{5\mu}(x)=J_{R\mu}(x)-J_{L\mu}(x)$. In the collapsing limit we find

$$\partial^{\mu}\langle\langle J^{(cs)}_{R\mu}(x)\rangle\rangle=\frac{1}{24\pi^2}\epsilon_{\mu\nu\lambda\rho}\,\partial^{\mu}V^{\nu}(x)\partial^{\lambda}V^{\rho}(x). \tag{88}$$

Similarly

$$\partial^\mu \langle\langle J_{L\mu}^{(cs)}(x)\rangle\rangle = -\frac{1}{24\pi^2}\epsilon_{\mu\nu\lambda\rho}\,\partial^\mu V^\nu(x)\partial^\lambda V^\rho(x). \tag{89}$$

These are the consistent right and left gauge anomalies—the label $^{(cs)}$ stands for consistent, to be distinguished from the covariant anomaly. As a matter of fact, application of the same chiral current splitting to the covariant anomaly of Equation (87) yields instead

$$\partial^\mu \langle\langle J_{R\mu}^{(cv)}(x)\rangle\rangle = \frac{1}{8\pi^2}\epsilon_{\mu\nu\lambda\rho}\,\partial^\mu V^\nu(x)\partial^\lambda V^\rho(x) \tag{90}$$

and

$$\partial^\mu \langle\langle J_{L\mu}^{(cv)}(x)\rangle\rangle = -\frac{1}{8\pi^2}\epsilon_{\mu\nu\lambda\rho}\,\partial^\mu V^\nu(x)\partial^\lambda V^\rho(x). \tag{91}$$

The label $^{(cv)}$ stands for covariant, and it is in order to tell apart these anomalies from the previous consistent ones. The two cases should not be confused: the consistent anomalies appears in the divergence of a current minimally coupled in the action to the vector potential V_μ. They represent the response of the effective action under a gauge transform of V_μ, which is supposed to propagate in the internal lines of the corresponding gauge theory. The covariant anomalies represent the response of the effective action under a gauge transform of the external axial current A_μ.

It goes without saying that, both for right and left currents in the collapsing limit, in the non-Abelian case the consistent anomaly takes the form (73), while the covariant one reads

$$\nabla \cdot \langle\langle J_R^a(x)\rangle\rangle = \frac{1}{32\pi^2}\epsilon_{\mu\nu\lambda\rho}\,\mathrm{Tr}\left(T^a F^{\mu\nu}(x)F^{\lambda\rho}(x)\right) \tag{92}$$

where $F_{\mu\nu}(x) = F^a_{\mu\nu}(x)T^a$ denotes the usual non-Abelian field strength. At first sight the above distinction between covariant and consistent anomalies for Weyl fermion may appear to be academic. After all, if a theory has a consistent anomaly it is ill-defined and the existence of a covariant anomaly may sound irrelevant. However this distinction becomes interesting in some non-Abelian cases since the non-Abelian consistent anomaly is proportional to the tensor d^{abc}. Now for most simple gauge groups (except $SU(N)$ for $N \geq 3$) this tensor vanishes identically. In such cases the consistent anomaly is absent and so the covariant anomaly becomes significant.

8. The Case of Majorana Fermions

As we have seen above, Majorana fermions are defined by the condition

$$\Psi = \widehat{\Psi}, \qquad \text{where} \qquad \widehat{\Psi} = \gamma_0 C \Psi^*. \tag{93}$$

Let $\psi_R = P_R\psi$ be a generic Weyl fermion. We have

$$P_R\psi_R = \psi_R \qquad\qquad P_L\widehat{\psi_R} = \widehat{\psi_R}$$

i.e., $\widehat{\psi_R}$ is left-handed. We have already remarked that $\psi_M = \psi_R + \widehat{\psi_R}$ is a Majorana fermion and any Majorana fermion can be represented in this way. Using this correspondence one can transfer the results for Weyl fermions to Majorana fermions[9]. The vector current is defined by $J^\mu_M = \bar\psi_M\gamma^\mu\psi_M$ and the axial current by $J^\mu_{5M} = \bar\psi_M\gamma^\mu\gamma_5\psi_M$. We can write

$$J^\mu_M(x) = \overline{\psi_R}(x)\gamma^\mu\psi_R(x) + \overline{\widehat{\psi_R}}(x)\gamma^\mu\widehat{\psi_R}(x) \equiv J^\mu_R(x) + J^\mu_L(x) \tag{94}$$

and

$$J^\mu_{5M}(x) = \overline{\psi_R}(x)\gamma^\mu\psi_R(x) - \overline{\widehat{\psi_R}}(x)\gamma^\mu\widehat{\psi_R}(x) \equiv J^\mu_R(x) - J^\mu_L(x). \tag{95}$$

Using (88) and (89) one concludes that, as far as the consistent anomaly is concerned,

$$\partial_\mu\langle\langle J^\mu_M(x)\rangle\rangle = 0 \tag{96}$$

This shows the consistency of our procedure, for one can show that, in general, $J_L(x) = -J_R(x)$, and $J_M(x) = 0$, as it should be for a Majorana fermion. On the other hand for the axial current we have

$$\partial_\mu\langle\langle J^\mu_{5M}(x)\rangle\rangle = \frac{1}{8\pi^2}\epsilon_{\mu\nu\lambda\rho}\,\partial^\mu V^\nu(x)\partial^\lambda V^\rho(x) \tag{97}$$

where the naïve sum has been divided by 2, because the two contributions come from the same degrees of freedom (which are half those of a Dirac fermion). From these results we see that, apart from the coefficient difference, the anomalies of a massless Majorana fermion are the same as those of a massless Dirac fermion. These obviously descend from the fact that both Dirac and Majorana fermions contain two opposite chiralities, at variance with a Weyl fermion, which is characterized by one single chirality.

9. Relation between Chiral and Trace Gauge Anomalies

There exists a strict relation between chiral gauge anomalies and trace anomalies in a theory of fermions coupled to a vector (and axial) gauge potential. This section is devoted to analysing this relation. When the background fields are not only V_μ and A_μ, but also a non-trivial tensor-axial metric $G_{\mu\nu} = g_{\mu\nu} + \gamma_5 f_{\mu\nu}$, see [17], the generating function must include two energy-momentum tensors, which in the flat-space limit take the Belifante–Rosenfeld symmetric form

$$T^{\mu\nu} = -\frac{i}{4}\left(\bar\psi\gamma^\mu\overset{\leftrightarrow}{\partial^\nu}\psi + \mu \leftrightarrow \nu\right), \tag{98}$$

and

$$T_5^{\mu\nu} = \frac{i}{4}\left(\bar\psi\gamma_5\gamma^\mu\overset{\leftrightarrow}{\partial^\nu}\psi + \mu \leftrightarrow \nu\right). \tag{99}$$

The quantities we are interested in here are, in particular, the 1-loop VEVs $\langle\langle T_{\mu\nu}(x)\rangle\rangle$ and $\langle\langle T_{5\mu\nu}(x)\rangle\rangle$ when $h_{\mu\nu} = f_{\mu\nu} = 0$: namely,

$$\begin{aligned}\langle\langle T_{\mu\nu}(x)\rangle\rangle &= \sum_{r,s=0}^{\infty}\frac{i^{r+s}}{2\,r!s!}\int\prod_{l=1}^{r}d^4x_l V^{\sigma_l}(x_l)\prod_{k=1}^{s}d^4y_k A^{\tau_k}(y_k) \\ &\times \langle 0|\mathcal{T}\,T_{\mu\nu}(x)J_{\sigma_1}(x_1)\dots J_{\sigma_r}(x_r)J_{5\tau_1}(y_1)\dots J_{5\tau_s}(y_s)|0\rangle\end{aligned} \tag{100}$$

[9] The converse is not true. It is impossible to reconstruct Weyl fermion anomalies from those of a massless Majorana fermion.

and

$$\langle\langle T_5^{\mu\nu}(x)\rangle\rangle = \sum_{r,s=0}^{\infty}\frac{i^{r+s}}{2\,r!s!}\int\prod_{l=1}^{r}d^4x_l V^{\sigma_l}(x_l)\prod_{k=1}^{s}d^4y_k A^{\tau_k}(y_k) \times \langle 0|\mathcal{T}\, T_5^{\mu\nu}(x)J_{\sigma_1}(x_1)\ldots J_{\sigma_r}(x_r)J_{5\tau_1}(y_1)\ldots J_{5\tau_s}(y_s)|0\rangle \tag{101}$$

of which we will compute the trace, i.e., contraction, over the indices μ and ν. Since we are interested in odd parity anomalies, the first nontrivial contributions come from the three-point correlators (i.e., $r+s=2$). Denoting by t, t_5 the traces of $T_{\mu\nu}, T_{5\mu\nu}$, the relevant correlators are $\langle t\, J\, J_5\rangle$ for (100), and $\langle t_5\, J\, J\rangle$, $\langle t_5\, J_5\, J_5\rangle$ for (101). We claim that they are simply related to $\langle\partial\cdot J\, J\, J_5\rangle$, $\langle\partial\cdot J_5\, J\, J\rangle$ and $\langle\partial\cdot J_5\, J_5\, J_5\rangle$, respectively.

We will also need

$$T_R^{\mu\nu}(x) = \tfrac{1}{2}\left[T^{\mu\nu}(x) + T_5^{\mu\nu}(x)\right]$$
$$T_L^{\mu\nu}(x) = \tfrac{1}{2}\left[T^{\mu\nu}(x) - T_5^{\mu\nu}(x)\right] \tag{102}$$

together with the one-loop 1-point function

$$\langle\langle T_{R,L}^{\mu\nu}(x)\rangle\rangle = \sum_{r=0}^{\infty}\frac{i^r}{r!}\int\prod_{l=1}^{r}d^4x_l V^{\sigma_l}(x_l)\langle 0|\mathcal{T}\, T_{R,L}^{\mu\nu}(x)\, J_{R,L\sigma_l}(x_l)|0\rangle. \tag{103}$$

9.1. Difference between Gauge and Trace Anomaly

Let us start from the case of the right-handed fermion. The correlator is, symbolically, $\langle t_R\, J_R\, J_R\rangle$, i.e., $\langle 0|\mathcal{T}\, T_{R\mu}^{\mu}(x)\, J_{R\lambda(y)}J_{R\rho(z)}|0\rangle$, its Fourier transform being given by

$$\widetilde{F}_{\mu\lambda\rho}^{(R)\mu}(k_1,k_2) = \frac{1}{4}\int\frac{d^4p}{(2\pi)^4}\mathrm{tr}\left[\frac{1}{\not p}\frac{1-\gamma_5}{2}\gamma_\lambda\frac{1}{\not p-\not k_1}\frac{1-\gamma_5}{2}\gamma_\rho\frac{1}{\not p-\not k_1-\not k_2}\frac{1-\gamma_5}{2}(2\not p-\not q)\right]. \tag{104}$$

The relevant Feynman diagram is shown in Figure 3.

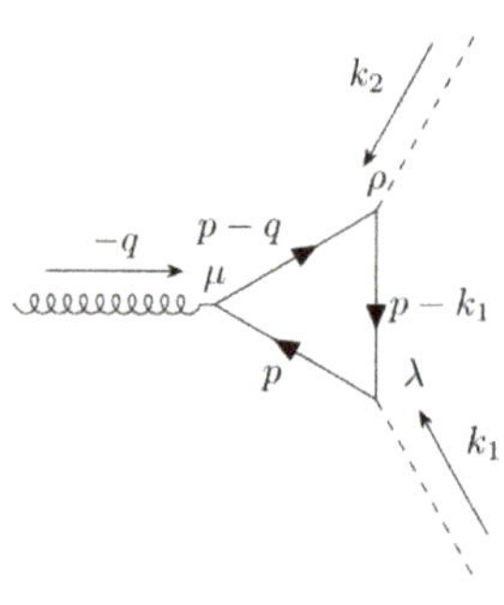

Figure 3. The Feynman diagram corresponding to $\widetilde{F}_{\mu\lambda\rho}^{(R)\mu}(k_1,k_2)$.

The difference with respect to the Fourier transform of $\langle \partial\cdot J_R\, J_R\, J_R\rangle$—see Equation (61)—apart from the factor $\frac{1}{4}$, is the $(2\not p - \not q)$ factor in the RHS, instead of $\not q$. The relevant difference is therefore twice

$$\Delta\widetilde{F}^{(R)\mu}_{\mu\lambda\rho}(k_1,k_2) = \frac{1}{4}\int\frac{d^4p}{(2\pi)^4}\mathrm{tr}\left[\frac{1}{\not p}\frac{1-\gamma_5}{2}\gamma_\lambda\frac{1}{\not p-\not k_1}\frac{1-\gamma_5}{2}\gamma_\rho\frac{1}{\not p-\not k_1-\not k_2}\frac{1-\gamma_5}{2}\not p\right] \tag{105}$$

$$= \frac{1}{4}\int\frac{d^4p d^\delta\ell}{(2\pi)^{4+\delta}}\mathrm{tr}\left[\frac{1}{\not p+\not \ell}\frac{1-\gamma_5}{2}\gamma_\lambda\frac{1}{\not p+\not \ell-\not k_1}\frac{1-\gamma_5}{2}\gamma_\rho\frac{1}{\not p+\not \ell-\not q}\frac{1-\gamma_5}{2}(\not p+\not \ell)\right]$$

$$= \frac{1}{4}\int\frac{d^4p d^\delta\ell}{(2\pi)^{4+\delta}}\frac{\mathrm{tr}\left[\gamma_\lambda(\not p-\not k_1)\gamma_\rho(\not p-\not q)\frac{1-\gamma_5}{2}\right]}{((p-k_1)^2-\ell^2)((p-q)^2-\ell^2)}. \tag{106}$$

We can now replace $p\to p+k_1$

$$\Delta\widetilde{F}^{(R)\mu}_{\mu\lambda\rho}(k_1,k_2) = \frac{i}{4}\int\frac{d^4p d^\delta\ell}{(2\pi)^{4+\delta}}\frac{\mathrm{tr}\left[\gamma_\lambda\not p\gamma_\rho(\not p-\not k_2)\frac{1-\gamma_5}{2}\right]}{(p^2-\ell^2)((p-k_2)^2-\ell^2)}. \tag{107}$$

The odd part vanishes by symmetry.

If we consider instead the amplitude for $\langle\partial\cdot J_5\, J\, J\rangle$, (80), the result does not change. In that case, for the odd part we get

$$\Delta\widetilde{F}^{(5)\mu}_{\mu\lambda\rho}(k_1,k_2) \sim \int\frac{d^4p d^\delta\ell}{(2\pi)^{4+\delta}}\mathrm{tr}\left(\gamma_\lambda\frac{\not p+\not \ell-\not k_1}{(p-k_1)^2-\ell^2}\gamma_\rho\frac{\not p+\not \ell-\not q}{(p-q)^2-\ell^2}\gamma_5\right)$$

$$= \int\frac{d^4p d^\delta\ell}{(2\pi)^{4+\delta}}\frac{(-\ell^2\mathrm{tr}(\gamma_\lambda\gamma_\rho\gamma_5)+\mathrm{tr}\left(\gamma_\lambda(\not p-\not k_1)\gamma_\rho(\not p-\not q)\gamma_5\right))}{((p-k_1)^2-\ell^2)((p-q)^2-\ell^2)}. \tag{108}$$

The first term in the numerator vanishes. The rest can be rewritten as

$$\Delta\widetilde{F}^{(5)\mu}_{\mu\lambda\rho}(k_1,k_2) \sim \int\frac{d^4p d^\delta\ell}{(2\pi)^{4+\delta}}\frac{\mathrm{tr}\left(\gamma_\lambda\not p\gamma_\rho(\not p-\not k_2)\gamma_5\right)}{(p^2-\ell^2)((p-k_2)^2-\ell^2)} = 0 \tag{109}$$

for the same reason as above. In the same way one can easily prove that

$$\Delta\widetilde{F}^{(5')\mu}_{\mu\lambda\rho}(k_1,k_2) = \Delta\widetilde{F}^{(5'')\mu}_{\mu\lambda\rho}(k_1,k_2) = 0. \tag{110}$$

In conclusion, the amplitude for the chiral anomalies and those for the trace anomalies due to couplings with gauge fields are rigidly related, the corresponding coefficients exhibiting a fixed ratio, i.e., the former are minus four times the latter.

9.2. Trace Anomalies Due to a Gauge Field

Using the above results, which say that the difference between $(-4)\times$[the divergence] and the trace anomaly is null, we can immediately deduce the corresponding consistent gauge trace anomalies, using (100), (101) and (103), viz.,

$$\langle\!\langle T^{(cs)\mu}_{R\mu}(x)\rangle\!\rangle = -\frac{1}{48\pi^2}\epsilon_{\mu\nu\lambda\rho}\partial^\mu V^\nu(x)\partial^\lambda V^\rho(x). \tag{111}$$

As for $T_{L\mu}(x)$, it carries the consistent anomaly

$$\langle\!\langle T^{(cs)\mu}_{L\mu}(x)\rangle\!\rangle = \frac{1}{48\pi^2}\epsilon_{\mu\nu\lambda\rho}\partial^\mu V^\nu(x)\partial^\lambda V^\rho(x). \tag{112}$$

On the other hand, in the $V-A$ framework we find

$$\langle\!\langle T^\mu_\mu(x)\rangle\!\rangle = 0 \tag{113}$$

and

$$\partial_\mu \langle\langle J_5^\mu(x) \rangle\rangle = -\frac{1}{16\pi^2}\epsilon_{\mu\nu\lambda\rho}\left(\partial^\mu V^\nu(x)\partial^\lambda V^\rho(x) + \frac{1}{3}\partial^\mu A^\nu(x)\partial^\lambda A^\rho(x)\right). \tag{114}$$

From (114) we can derive the covariant chiral anomaly for a Dirac fermion by setting $A_\mu = 0$, then

$$\langle\langle T_{5\mu}^{(cv)\mu}(x) \rangle\rangle = \frac{1}{16\pi^2}\epsilon_{\mu\nu\lambda\rho}\,\partial^\mu V^\nu(x)\partial^\lambda V^\rho(x). \tag{115}$$

From this we can derive the covariant (invariant) trace anomaly for a right-handed

$$\langle\langle T_{R\mu}^{(cv)\mu}(x) \rangle\rangle = -\frac{1}{16\pi^2}\epsilon_{\mu\nu\lambda\rho}\,\partial^\mu V^\nu(x)\partial^\lambda V^\rho(x) \tag{116}$$

and left-handed Weyl fermion

$$\langle\langle T_{L\mu}^{(cv)\mu}(x) \rangle\rangle = \frac{1}{16\pi^2}\epsilon_{\mu\nu\lambda\rho}\,\partial^\mu V^\nu(x)\partial^\lambda V^\rho(x). \tag{117}$$

9.3. Gauge Anomalies and Diffeomorphisms

In this review we have not considered diffeomorphisms. Nevertheless a devil's accountant could argue that there might be violation of diffeomorphism invariance in a fermionic system coupled to gauge fields, due to the presence of the gauge fields themselves. In order to see this one has to consider three point correlators involving the divergence of the energy momentum tensor and two currents. More precisely, odd parity anomalies could appear in the following amplitudes: $\langle \partial\cdot T_R\, J_R\, J_R \rangle$, in the right-handed fermion case, or $\langle \partial\cdot T_5\, J\, J \rangle$, $\langle \partial\cdot T\, J_5\, J \rangle$,$\langle \partial\cdot T_5\, J_5\, J_5 \rangle$ in the $V - A$ case. They can be computed with the same methods as above, and here, for brevity, we limit ourselves to record the final results: they all vanish.

10. Conclusions

The purpose of this review was to highlight some subtle aspects of the physics of Weyl fermions, as opposed in particular to massless Majorana spinors. To this end we have decided not to resort to powerful non-perturbative methods, like the Seeley–Schwinger–DeWitt method, which would require a demanding introduction. Rather, we have used the simple Feynman diagram technique. In doing so, we have focused on two aims. The first one is to justify the method of computing the effective action for a Weyl fermion coupled to gauge potentials, which requires the presence of free fermions of opposite chirality, in such a way as to produce the effective kinetic operator of Equation (30). We have shown that, notwithstanding the presence of fermions of both chiralities, no mass term can arise as a consequence of quantum corrections. As a by-product we were led to the conclusion that, while the Pauli–Villars regularisation is a perfectly available and useful tool for perturbative calculations, it does not fit at all in the case of non-perturbative heat kernel-like methods.

Our second aim was to compute all the anomalies (trace and chiral) of Weyl, Dirac and Majorana fermions coupled to gauge potentials. The calculations are actually standard, but, once juxtaposed, they reveal a perhaps previously unremarked property: the chiral and trace anomalies due to a gauge background are rigidly linked.

Author Contributions: All author contribuite equally. All authors have read and agreed to the published version of the manuscript.

Funding: This research received no external funding.

Acknowledgments: S.Z. thanks SISSA for hospitality while this work was being completed. L.B. would like to thank his former collaborators A. Andrianov, M. Cvitan, P. Dominis-Prester, A. Duarte Pereira, S. Giaccari and T. Štemberga, with whom part of the material of this paper was elaborated or discussed.

Conflicts of Interest: The authors declare no conflict of interest

Appendix A. Spin-States for Weyl Spinor Fields

In this appendix we aim to summarize the explicit form of the spin-states and plane wave functions for Weyl spinor fields in the more general and useful four component formalism. In order to realise a basis of spin states for a Weyl spinor field, we have to search among states of opposite chirality and frequency. Then, in so doing, we will be able to have at hand a complete and orthogonal quartet of spin-states for the 4D massless bispinor space. After setting $p^{\mu} = (|\mathbf{p}|, \mathbf{p}) = (\wp, p_x, p_y, p_z)$, we have

$$u_-(\mathbf{p}) = \frac{1}{\sqrt{\wp - p_z}} \begin{pmatrix} \wp - p_z \\ -p_x - ip_y \\ 0 \\ 0 \end{pmatrix}$$

The above spin-state is a positive frequency solution of the Weyl field equation and exhibits negative chirality: namely,

$$\not{p}\, u_-(\mathbf{p}) = 0 \qquad (\gamma_5 + 1)\, u_-(\mathbf{p}) = 0$$

Next we set

$$u_+(\mathbf{p}) = \frac{-1}{\sqrt{\wp + p_z}} \begin{pmatrix} 0 \\ 0 \\ \wp + p_z \\ p_x + ip_y \end{pmatrix}$$

which satisfies by construction

$$\not{p}\, u_+(\mathbf{p}) = 0 \qquad (\gamma_5 - 1)\, u_+(\mathbf{p}) = 0$$

Quite analogously, we can build up as well a pair of negative frequency spin-states of momentum $\mathbf{p}$ and both opposite chiralities: namely,

$$\not{\tilde{p}}\, v_{\mp}(\mathbf{p}) = 0 \qquad (\gamma_5 \pm 1)\, v_{\mp}(\mathbf{p}) = 0$$

where $\tilde{p}^{\mu} = p_{\mu}$. We find

$$v_-(\mathbf{p}) = \frac{1}{\sqrt{\wp - p_z}} \begin{pmatrix} p_x - ip_y \\ \wp - p_z \\ 0 \\ 0 \end{pmatrix}$$

together with

$$v_+(\mathbf{p}) = \frac{1}{\sqrt{\wp + p_z}} \begin{pmatrix} 0 \\ 0 \\ -p_x + ip_y \\ \wp + p_z \end{pmatrix}$$

It turns out that the above defined quartet of massless and chiral bispinor spin-states does fulfil orthogonality and closure relations. As a matter of fact, spin-states of opposite chirality and/or opposite frequency are orthogonal, as it does, and furthermore we get

$$(2\wp)^{-1}[\, u_{\pm}(\mathbf{p}) \otimes u_{\pm}^{\dagger}(\mathbf{p}) + v_{\pm}(\mathbf{p}) \otimes v_{\pm}^{\dagger}(\mathbf{p})\,] = \mathbb{P}_{\pm}$$

with of course $\mathbb{P}_{\pm} = \frac{1}{2}(\,\mathbb{I} \pm \gamma_5)$ which represent the closure or completeness relations for the massless and chiral bispinor spin-states basis quartet. The corresponding plane-wave functions, which are normal-mode solutions of the massless Dirac equation, read

$$\begin{cases} u_{\pm,\mathbf{p}}(x) = [\,(2\pi)^3 2\wp\,]^{-\frac{1}{2}} u_{\pm}(\mathbf{p})\, e^{-i\wp t + i\mathbf{p}\cdot\mathbf{x}} \\ v_{\pm,\mathbf{p}}(x) = [\,(2\pi)^3 2\wp\,]^{-\frac{1}{2}}\, v_{\pm}(-\,\mathbf{p})\, e^{i\wp t - i\mathbf{p}\cdot\mathbf{x}} \end{cases}$$

and fulfil in turn orthonormality and closure relation with respect to the usual Poincaré invariant inner product. It is crucial to gather, for a better understanding of the matter, that for any given frequency and chirality the spin-states and wave-functions of particle and anti-particle enjoy opposite wave vectors, i.e., opposite helicity.

Appendix B. Consistent Gauge Anomaly with PV Regularisation

To implement a PV regularisation, we replace $\widetilde{F}^{(R)}_{\mu\nu\lambda}(k_1,k_2)$ with

$$\begin{aligned} \widetilde{F}^{(R)}_{\mu\nu\lambda}(k_1,k_2) \;=\; & \int \frac{d^4p}{(2\pi)^4} \mathrm{tr}\left\{ \frac{1}{\not p + m}\frac{1-\gamma_5}{2}\gamma_\nu \frac{1}{\not p - \not k_1 + m}\frac{1-\gamma_5}{2}\gamma_\lambda \frac{1}{\not p - \not q + m}\frac{1-\gamma_5}{2}\gamma_\mu \right. \\ & \left. - \frac{1}{\not p + M}\frac{1-\gamma_5}{2}\gamma_\nu \frac{1}{\not p - \not k_1 + M}\frac{1-\gamma_5}{2}\gamma_\lambda \frac{1}{\not p - \not q + M}\frac{1-\gamma_5}{2}\gamma_\mu \right\} \end{aligned} \tag{A1}$$

m and M are IR and UV regulators, respectively, and tr is the trace of gamma matrices.

Contracting with q^μ and working out the traces one gets

$$q^\mu \widetilde{F}^{(R)}_{\mu\nu\lambda}(k_1,k_2) \;=\; -2i\epsilon_{\mu\nu\rho\lambda} \int \frac{d^4p}{(2\pi)^4} \left(2p\cdot q\, p^\mu - p^2 q^\mu - q^2 p^\mu\right)(p-k_1)^\rho \left(\frac{1}{\Delta_{m^2}} - \frac{1}{\Delta_{M^2}}\right) \tag{A2}$$

where

$$\begin{aligned} \Delta_{m^2} &= (p^2-m^2)((p-k_1)^2-m^2)((p-q)^2-m^2), \\ \Delta_{M^2} &= (p^2-M^2)((p-k_1)^2-M^2)((p-q)^2-M^2) \end{aligned}$$

For later use we introduce also

$$\Omega_{m^2} = ((p-k_1)^2-m^2)((p-q)^2-m^2), \qquad \Lambda_{m^2} = (p^2-m^2)((p-k_1)^2-m^2), \tag{A3}$$

$$\Omega_{M^2} = ((p-k_1)^2-M^2)((p-q)^2-M^2), \qquad \Lambda_{M^2} = (p^2-M^2)((p-k_1)^2-M^2). \tag{A4}$$

Now all the integrals are convergent because the divergent terms have been subtracted away. Let us proceed

$$\begin{aligned} & q^\mu \widetilde{F}^{(R)}_{\mu\nu\lambda}(k_1,k_2) = -2i\epsilon_{\mu\nu\rho\lambda} \int \frac{d^4p}{(2\pi)^4} \left\{ \left[\frac{-k_2^\mu (p-k_1)^\rho}{\Omega_{m^2}\Delta_{M^2}} + \frac{p^\mu k_1^\rho}{\Lambda_{m^2}\Delta_{M^2}} \right] \right. \\ & \left(m^6 - M^6 + \left(M^4 - m^4\right)\left(p^2 + (p-k_1)^2 + (p-q)^2\right) \right. \\ & \left. + \left(m^2 - M^2\right)\left((p-k_1)^2(p-q)^2 + (p-k_1)^2 p^2 + p^2(p-q)^2\right)\right) \\ & \left. + m^2\left(p^\mu k_1^\rho - q^\mu (p-k_1)^\rho\right)\left(\frac{1}{\Delta_{m^2}} - \frac{1}{\Delta_{M^2}}\right)\right\} \end{aligned} \tag{A5}$$

The last line does not contribute, for the integrals converge (separately) and give a finite result, but since they are multiplied by m^2 they vanish in the limit $m \to 0$. Therefore, the last line can be dropped.

Now the strategy consists in simplifying separately each monomials in the numerator with a corresponding term in the denominator. For instance, if in a term of order M^* there is the ratio $p^2/(p^2-m^2)$, write p^2 as $p^2-m^2+m^2$. The p^2-m^2 can be simplified with a corresponding term in the denominator. If p^2-m^2 in the denominator is missing, there will be p^2-M^2. We write p^2 as $p^2-M^2+M^2$, and p^2+M^2 can be simplified, while the term proportional to M^2 remains and contributes to the term of order M^{*+2}. Proceed in the same way also with $(p-q)^2$ and $(p-k_1)^2$. Many terms (such as those of order M^6) cancel out. What remains is

$$
\begin{aligned}
q^\mu \widetilde{F}^{(R)}_{\mu\nu\lambda}(k_1,k_2) &= -2i\epsilon_{\mu\nu\rho\lambda}\int \frac{d^4p}{(2\pi)^4}\Bigg\{ \\
&p^\mu k_1^\rho\left[\frac{(M^2-m^2)^2}{\Lambda_{m^2}\Lambda_{M^2}} - \frac{(M^2-m^2)}{\Lambda_{M^2}}\left(\frac{1}{p^2-m^2}+\frac{1}{(p-k_1)^2-m^2}\right) - \frac{M^2-m^2}{\Delta_{M^2}}\right] \\
&-k_2^\mu(p-k_1)^\rho\left[\frac{(M^2-m^2)^2}{\Omega_{m^2}\Omega_{M^2}} - \frac{(M^2-m^2)}{\Omega_{M^2}}\left(\frac{1}{(p-k_1)^2-m^2}+\frac{1}{(p-q)^2-m^2}\right) - \frac{M^2-m^2}{\Delta_{M^2}}\right]\Bigg\}
\end{aligned}
\tag{A6}
$$

It is easy too verify that, after introducing the relevant Feynman parameters, most of the terms vanish either because there is only one p in the numerator or because of the anti-symmetry of the ϵ tensor. Only the last term in each line remains, so that:

$$
q^\mu \widetilde{F}^{(R)}_{\mu\nu\lambda}(k_1,k_2) = 2i\,M^2\epsilon_{\mu\nu\rho\lambda}\int \frac{d^4p}{(2\pi)^4}\frac{p^\mu k_1^\rho - k_2^\mu(p-k_1)^\rho}{\Delta_{M^2}}
\tag{A7}
$$

Next we introduce two Feynman parameters x and y, shift p like in Section 6.1 and make a Wick rotation on the momenta. Then, (A6) becomes

$$
\begin{aligned}
q^\mu \widetilde{F}^{(R)}_{\mu\nu\lambda}(k_1,k_2) &= -4\,M^2\epsilon_{\mu\nu\rho\lambda}\int_0^1 dx\int_0^{1-x}dy\int\frac{d^4p}{(2\pi)^4}\frac{(1-x)k_1^\mu k_2^\rho}{(p^2+M^2+A(x,y))^3} \\
&= -\frac{1}{8\pi^2}\,M^2\epsilon_{\mu\nu\rho\lambda}\int_0^1 dx\int_0^{1-x}dy\,\frac{(1-x)k_1^\mu k_2^\rho}{M^2+A(x,y)} \\
&= -\frac{1}{24\pi^2}\epsilon_{\mu\nu\rho\lambda}k_1^\mu k_2^\rho
\end{aligned}
\tag{A8}
$$

Adding the cross term we get

$$
q^\mu \widetilde{T}^{(R)}_{\mu\nu\lambda}(k_1,k_2) = -\frac{1}{12\pi^2}\epsilon_{\mu\nu\rho\lambda}k_1^\mu k_2^\rho
\tag{A9}
$$

which is the same result as in Section 6.1.

Appendix C. Consistent and Covariant Gauge Anomalies

In this Appendix we briefly recall the difference between covariant and consistent anomalies. The reason why (87) is called covariant is self-evident, whereas why (73) is called consistent is not so straightforward.

Let us consider a generic gauge theory, with connection $V^a_\mu T^a$, valued in a Lie algebra $\mathfrak{g}$ with anti-Hermitean generators T^a , such that $[T^a,T^b]=f^{abc}T^c$. In the following it is convenient to use the more compact form notation and represent the connection as a one-form $\mathbf{V}=A^a_\mu T^a dx^\mu$, so that the gauge transformation becomes

$$
\delta_\lambda \mathbf{V} = \mathbf{d}\lambda + [\mathbf{V},\lambda]
\tag{A10}
$$

with $\lambda(x) = \lambda^a(x)T^a$ and $\mathbf{d} = dx^\mu \frac{\partial}{\partial x^\mu}$. The mathematical problem is better formulated if we promote the gauge parameter λ to an anticommuting field $c = c^a T$, the Faddeev–Popov ghost, and define the BRST transformation as

$$s\mathbf{V} = \mathbf{d}c + [\mathbf{V}, c], \qquad sc = -\frac{1}{2}[c, c] \tag{A11}$$

The operation s is nilpotent, $s^2 = 0$. We represent with the same symbol s the corresponding functional operator, i.e.,

$$s = \int d^d x \left(s\mathbf{V}^a(x) \frac{\partial}{\partial \mathbf{V}^a(x)} + sc^a(x) \frac{\partial}{\partial c^a(x)} \right) \tag{A12}$$

The Ward identity for the gauge (BRST) transformation is classically given by $s\,W[V] = 0$, but at the first quantum level one may find an anomaly

$$s\,W[V] = \hbar \mathcal{A}[V, c] \tag{A13}$$

the RHS is an integrated expression linear in c, and once the latter is removed, in 4D it becomes precisely the RHS of (73).

Due to the nilpotency of s the anomaly must satisfy

$$s\mathcal{A}[V, c] = 0 \tag{A14}$$

These are the famous Wess–Zumino consistency conditions (it is enough to functionally differentiate (A14) by c^a to cast them in the original form). $\mathcal{A}[V, c]$ is a cocycle of the cohomology defined by s. If there exists a local expression $\mathcal{C}[V]$ such that $\mathcal{A}[V, c] = s\mathcal{C}[V]$, the cocycle is a coboundary (trivial anomaly). If such $\mathcal{C}[V]$ does not exist, we have a true anomaly.

An easy way to verify the consistency conditions is via the descent equations. To construct the descent equations we start from a symmetric polynomial in the Lie algebra of order 3, $P_3(T^a, T^b, T^c)$, invariant under the adjoint transformation:

$$P_3([X, T^a], T^b, T^c) + \ldots + P_3(T^a, T^b, [X, T^c]) = 0 \tag{A15}$$

for any element X of the Lie algebra. In many cases these polynomials are symmetric traces of the generators in the corresponding representation

$$P_3(T^a, T^b, T^c) = Str(T^a T^b T^c) = d^{abc} \tag{A16}$$

(Str denotes the symmetric trace). With this one can construct the 6-form

$$\Delta_6(\mathbf{V}) = P_n(\mathbf{F}, \mathbf{F}, \ldots \mathbf{F}) \tag{A17}$$

where $\mathbf{F} = \mathbf{d}\mathbf{V} + \frac{1}{2}[\mathbf{V}, \mathbf{V}]$. It is easy to prove that

$$P_3(\mathbf{F}, \mathbf{F}, \ldots \mathbf{F}) = \mathbf{d}\left(n \int_0^1 dt\, P_n(\mathbf{V}, \mathbf{F}_t, \ldots, \mathbf{F}_t) \right) = \mathbf{d}\Delta^{(0)}_{2n-1}(\mathbf{V}) \tag{A18}$$

where we have introduced the symbols $\mathbf{V}_t = t\mathbf{V}$ and its curvature $\mathbf{F}_t = \mathbf{d}\mathbf{V}_t + \frac{1}{2}[\mathbf{V}_t, \mathbf{V}_t]$, where $0 \leq t \leq 1$. In the above expressions the product of forms is understood to be the exterior product. To prove

Equation (A18) one uses in an essential way the symmetry of P_3 and the graded communtativity of the exterior product of forms. Equation (A18) is the first of a sequence of equations that can be proven

$$\Delta_6(\mathbf{V}) - \mathbf{d}\Delta_5^{(0)}(\mathbf{V}) = 0 \tag{A19}$$

$$s\Delta_5^{(0)}(\mathbf{V}) - \mathbf{d}\Delta_4^{(1)}(\mathbf{V},c) = 0 \tag{A20}$$

$$s\Delta_4^{(1)}(\mathbf{V},c) - \mathbf{d}\Delta_3^{(2)}(\mathbf{V},c) = 0 \tag{A21}$$

$$\ldots\ldots$$

$$s\Delta_0^{(5)}(c) = 0 \tag{A22}$$

All the expressions $\Delta_k^{(p)}(V,c)$ are polynomials of $\mathbf{d}, c$ and $\mathbf{V}$. The lower index k is the form order and the upper one p is the ghost number, i.e., the number of c factors. The last polynomial $\Delta_0^{(5)}(c)$ is a 0-form and clearly a function only of c. All these polynomials have explicit compact form. They were originally introduced by S.S. Chern [20], in the geometric context of a principal fibre bundle, which gives them full meaning. Here we limit ourselves to a formal application, which however can be proven to lead to correct results.

The RHS contains the general expression of the consistent gauge anomaly in 4 dimensions, for, integrating (3.53) over space-time, one gets

$$\begin{aligned} s\,\boldsymbol{\mathcal{A}}[c,\mathbf{V}] &= 0 \\ \boldsymbol{\mathcal{A}}[c,\mathbf{V}] &= \int d^dx\,\Delta_4^1(c,\mathbf{V}), \qquad \text{where} \\ \Delta_4^1(c,\mathbf{V}) &= 6\int_0^1 dt(1-t)P_3(\mathbf{d}c,\mathbf{V},\mathbf{F}_t) \end{aligned} \tag{A23}$$

One can verify that $\boldsymbol{\mathcal{A}}[c,\mathbf{V}]$ is the anomaly (73), saturated with c^a and integrated over space-time, up to an overall numerical coefficient.

To verify that the anomaly (73) is non-trivial one can use a *reductio ad absurdum* argument. Let us suppose that (A23) is trivial. Then using repeatedly the local Poincaré lemma one can prove that there must exist a 0-form $C_0^{(2n-2)}(c)$ such that

$$\Delta_0^{(5)}(c) = s\,C_0^{(4)}(c) \tag{A24}$$

However this is impossible, for the expression for $\Delta_0^{(5)}(c)$ is

$$\Delta_0^{(5)}(c) \sim P_n(c,[c,c]_+,\ldots,[c,c]_+) \tag{A25}$$

and the only possibility for $C_0^{(4)}(c)$ to satisfy (A24) is to have the form

$$C_0^{(4)}(c) \sim P_n(c,c,[c,c]_+) \tag{A26}$$

which, however, vanishes due to the symmetry of P_3 and the anticommutativity of c.

It is evident that in the Abelian case the above construction becomes trivial, and consistent and covariant gauge anomalies collapse to the same expressions. The only difference is the numerical factor in front of them.

References

1. Weyl, H.K.H. *Gruppentheorie und Quantenmechanik*; Hirzel: Leipzig, Germany, 1928; Chapter VIII, p. 288.
2. Gel'fand, I.M.; Naimark, M.A. Unitary representations of the Lorentz group. *Izv. Akad. Nauk. SSSR Ser. Mat.* **1947**, *11*, 411–504
3. Wigner, E.P. *Gruppentheorie und ihre Anwendung auf die Quantenmechanik der Atomspektrum*; Fredrick Vieweg und Sohn: Braunschweig, Germany, 1931; pp. 251–254.
4. Wigner, E.P. *Group Theory and Its Application to the Quantum Theory of Atomic Spectra*; Academic Press Inc.: New York, NY, USA, 1959; pp. 233–236.
5. Racah, G. On the symmetry of particles and antiparticles. *Nuovo Cimento* **1936**, *14*, 322. [CrossRef]
6. Schwinger, J. On gauge invariance and vacuum polarization. *Phys. Rev.* **1951**, *82*, 664–679. [CrossRef]
7. DeWitt, B.S. *The Dynamical Theory of Groups and Fields*; Gordon & Breach: New York, NY, USA, 1965.
8. Seeley, R.T. Complex powers of an elliptic operator. *Electron. Res. Announc. Am. Math. Soc.* **1967**, *10*, 288–307
9. Seeley, R.T. The resolvent of an elliptic boundary value problem. *Am. J. Math.* **1969**, *91*, 889–920. [CrossRef]
10. Hawking, S.W. Zeta function regularization of path integrals in curved space-time. *Commun. Math. Phys.* **1977**, *55*, 133–162. [CrossRef]
11. Cornwell, J.F. *Group Theory in Physics*; Academic Press: London, UK, 1984; Chapter 17, Volume II.
12. Bertlmann, R.A. *Anomalies in Quantum Field Theory*; Oxford University Press: Oxford, UK, 1996.
13. Kirsten, K. Heat kernel asymptotics: More special case calculations. *Nucl. Phys. B (Proc. Suppl.)* **2002**, *104*, 119–126. [CrossRef]
14. Vassilevich, D.V. Heat kernel expansion: User's manual. *Phys. Rep.* **2003**, *388*, 279–360. [CrossRef]
15. DeWitt, B.S. *Global Approach to Quantum Field Theory*; Oxford University Press: Oxford, UK, 2003; Volumes I and II.
16. Pal, P.B. Dirac, Majorana and Weyl fermions. *arXiv* **2011**, arXiv:1006.1718.
17. Bonora, L.; Cvitan, M.; Prester, P.D.; Pereira, A.D.; Giaccari, S.; Štemberga, T. Axial gravity, massless fermions and trace anomalies. *Eur. Phys. J. C* **2017**, *77*, 511. [CrossRef]
18. Bonora, L.; Soldati, R. On the trace anomaly for Weyl fermions. *arXiv* **2019**, arXiv:1909.11991.
19. Bardeen, W.A. Anomalous Ward Identities in Spinor Field Theories. *Phys. Rev.* **1969**, *184*, 1848. [CrossRef]
20. Chern, S.S. *Complex Manifolds without Potential Theory*; Springer: Berlin, Germany, 1969.

MDPI
St. Alban-Anlage 66
4052 Basel
Switzerland
Tel. +41 61 683 77 34
Fax +41 61 302 89 18
www.mdpi.com

Universe Editorial Office
E-mail: universe@mdpi.com
www.mdpi.com/journal/universe

www.ingramcontent.com/pod-product-compliance
Lightning Source LLC
Chambersburg PA
CBHW062035220926
51850CB00010B/2551
9783036521978